CW00400378

Relativity Explained
(without maths)
Behind the curtain of reality

Independently published
ISBN 9798509255717   (paperback, colour)
ISBN 9798523512100   (paperback, black & white)
ASIN B01LSR1T2Q   (eBook)

For my parents Viviane & Henri
who made me curious about the world

The most beautiful thing we can experience is the mysterious. It is the source of all true art and science. He to whom the emotion is a stranger, who can no longer pause to wonder and stand wrapped in awe, is as good as dead; his eyes are closed.

Albert Einstein

Relativity Explained

(without maths)

Behind the curtain of reality

Contents

PREFACE

THE AIM OF THIS BOOK

This book aims to explain the revelations of relativity and, to a lesser extent, of quantum mechanics. It also aims to explain how the universe actually works (as far as is known) and reveal what lies behind the curtain of our perceived reality.

THIS SMALL BOOK IS FOR YOU IF ......

... any of the following applies. You ...

- want to understand relativity – without any mathematics,
- feel that books with appealing titles such as 'Relativity made easy' have been written by brilliant minds which understand their subject too well, but have forgotten how hard it can be to grasp in the first place*,
- have heard the basic ideas that: nothing travels faster than light; something called 'spacetime' is four-dimensional; length, time and mass change as the speed of light is approached; and that $E=mc^2$ – but you struggle to understand if or why these things are true or how they are all connected,
- feel that without some understanding of relativity its bold practical and philosophical implications lie even further beyond your disbelieving grasp constraining your attempts to stare in amazement at our re-envisioned universe,
- feel that surely there must be a way to communicate a convincing explanation of relativity to a layman over just a cup of coffee, or
- just want to peak behind the curtain of reality.

* My advantage is that: (a) I am not that clever, (b) I wrote this book whilst still on my own journey of discovery, and (c) this book has been expertly reviewed (see *Acknowledgments* on page 131).

Why is it Useful to Understand Relativity?
There are three reasons.

Knowledge for its own sake
The development of human society relies on the cumulative development of ideas. When those ideas are wrong we take wrong turns. When they are right we advance. The more we know the better we become.

Building the future
Without Sir Isaac Newton's equations of motion swathes of useful technology would not have been possible: from aerospace to biomechanics to construction and beyond. Engineers informed and inspired by science are better placed to build a better world.

Who we are and what sort of world we live in
There is no necessity to understand everything but some things are more important than others. High on that list is a sense of who we are individually and as a species. It is a subject best informed by facts.

We learned 400 years ago that Man was not the centre of the universe and only 100 years ago that there was much more to the universe than just the visible sky; indeed, that we may not be alone in the universe. Such discoveries have changed our ideas of who we are.

It wasn't wrong for early Man to think himself the centre of the universe, it fitted the facts then available. But it is wrong that beliefs based on primitive ideas prove unshakeable, despite contradictory facts, for far too long. Through relativity our awakening continues.

Why Else Might You Want to Understand Relativity?
Of all the reasons given on the previous page the one that mattered most to me was to gain some comprehension of our universe and the "bold practical and philosophical implications" of 20th century physics.

Put another way, why on earth wouldn't I wish to correct my inaccurate understanding of our perceived reality?

Chapter

TO PEEK BEHIND THE CURTAIN OF REALITY

BEHIND THE CURTAIN

The magician has finished her act, you are amazed. She invites you backstage, behind the curtain, where she will reveal the secret. Why wouldn't you follow her?

We wander through our lives certain in our belief that we occupy space and that time passes. Much as we occupy a room and watch the hands of the clock on the wall mark the passage of time.

Enter Einstein, the magician's assistant, who reveals that this is an illusion – that we have all been fooled. He entices us backstage saying that everything we see and touch, every force we experience, everything including the illusions of space and time themselves are all woven from the same thread, that our reality is not the reality.

And with that he disappears behind the curtain.

Why wouldn't we follow him?

A child-like question – A universe-shattering answer
Albert Einstein was 16 years old when he first wondered how the universe would look if he could ride on a beam of light. This is the story of how that simple question eventually revealed a startlingly different universe to the one we believe we inhabit.

Time, not what it seems

'When?' 'How long?' 'How old?' Time in its many guises is central to every human thought and action. We imagine we know what time is until, that is, we are required to explain precisely what it is. At that point we realise that we've never seen it nor do we fully comprehend what it is.

Despite this uncertainty it still comes as a shock to discover that relativity and quantum mechanics, the twin breakthroughs of 20th century physics, at the very least call into question any intuitive notions we may have about what time is. Of the many surprises which lie behind the curtain of reality this is one of the greatest.

Einstein's legacy

Einstein made a number of predictions based on relativity theory. They all proved correct. The last of these, that waves of gravity cause space and time and indeed matter itself to ripple, was proved in 2016.

The equipment which detected these 'gravitational waves' opens a new field of astronomical observation. Over the last 100 years ever more powerful telescopes using light or radio waves have revealed exotic cosmological objects scarcely imagined before. And now gravitational wave 'telescopes' will do likewise. This is Einstein's final gift, enabling us to peer even further behind the curtain of reality.

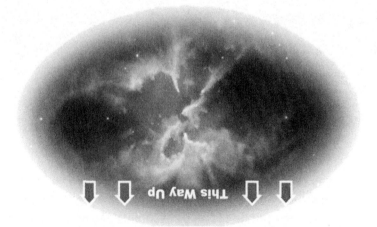

The universe is not as it seems

A SNEAK PREVIEW

Relativity makes many incredible claims so it is perhaps helpful here to provide a very high level overview, a sneak preview. The rest of this book will try to explain what lies behind these claims.

For many people discovering that space contracts and time slows if we travel very fast is so incredible that it blocks all further attempts to grasp relativity. Our minds hold fast to their own intuitions: that space is an unbending container, and that time beats to a constant rhythm. However these intuitions, for that is all they are, are the product of our limited earth-bound experiences and ideas.

The earth as seen by NASA's Voyager 1 spacecraft
as it left our solar system in 1990.
'A mote of dust suspended in a sunbeam,' as Carl Sagan described it.

Our earth is, however, a far far far smaller fraction of the universe than a grain of sand is in relation to the whole earth. And yet we mere humans seem to believe that the workings of the whole universe are an extension of our earth-bound intuitions. Why should that be so?

Surely it's equally plausible that our intuitions are misguided *because* they're earth-bound; that words like 'space' and 'time' to which we so firmly attach one meaning might just as easily mean something different. In actual fact simply by using our imaginations we can glimpse these different possibilities.

3

Thinking up space and time

Imagine floating in empty space. You have no equipment: no light, no watch, no ruler, nothing – just you and space. The nearest galaxies are far away, not even pinpricks of light, invisible to the naked eye. (The average distance between galaxies in our universe is around 10 million million million kilometres – i.e. much more than 10 billion times the distance from the earth to the sun.) Wherever you look you see only velvety black darkness. Now ask yourself a few questions.

How empty is the space around you? Presumably, you are 'in' something that supports you floating. Maybe there is something even emptier than empty space, let's call it a 'complete void'. 'Empty space' might then be some sort of supporting medium which, albeit invisible to us, occupies this 'complete void'.

Which way is up? We tend to imagine that things like direction are features of space. But, as you look around there is no trace of such features. And, while on earth we use man-made rulers to measure space, out here the seemingly endless velvety blackness blocks our sense of distance, scale or direction.

Are you stationary or drifting? If drifting, how fast? In the vast blackness, without any defining features, there is no means of knowing. You realise that movement and speed make sense only by comparison to other objects.

How quickly time is passing is equally unknowable. Even if you had a watch how would you know it wasn't ticking too fast or too slowly?

The problem of time goes much deeper. As you look around there is no day and night, nothing moves. There are no moment to moment changes in the scenery around you. At best, time has lost all purpose. It might even be that time as we know it is man-made, invented simply to record events as they happen around us.

You begin to understand that the way we *describe and measure* space and time are certainly human inventions. And you even start to wonder if we earth-bound humans *dreamt up* our very own ideas of space and time in the first place.

Perhaps the words 'space' and 'time' do mean something different after all.

Thinking up a different space and time
What might space be? We think of it as an unbending container that holds the stuff of the universe. But we can't *know* if that's correct. The only thing we know for certain is that something we call 'space' separates events from one another: so that they don't all happen in the same place.

What might time be? We tend to think of time as the constant rhythm of the universe. But we can't *know* if that's correct. The only thing we know for certain is that something we call 'time' separates events from one another: so that they don't all happen at the same time.

With these stripped down descriptions, space and time appear more similar than we tend to imagine. After all, an event at a point in space is also an event at a point in time.

Space and time have something else in common, something not obvious at first glance, they both 'flow'. We experience the passage of time: from 'before' through 'now' to 'after'. Likewise, we experience our passage through space: from 'behind' through 'here' to 'ahead'.

Trading space and time
These similarities point to a profound change in our earth-bound conceptions of space and time. They are not totally separate entities, but in fact are related to each other, each a part of something physicists call 'spacetime'.

So, when we think we are moving through space we are actually moving through spacetime. In other words, not just through space but also through time – such that any increase in our speed through space will be traded off at the expense of our speed through time. As a result, time will appear to have slowed.

Furthermore, since we define length as 'the distance between two points *measured at the same moment in time*' it suggests that we will have different perceptions of length also.

But for two people not moving relative to each other, i.e. no speed difference between them, they will each have the same perception of the passing of time and the same perception of length will also.

The medium of spacetime
If our experience of space and time depends on the motion of objects travelling through spacetime, that implies that spacetime is not just a container for everything in the universe but is as much a part of the universe as you or me.

Much like objects suspended in water, the universe comprises objects suspended in spacetime. Objects feel the pressure of spacetime around them, and they also exert an influence of their own, a back pressure, which distorts spacetime around the object.

Herein lies the reason for our experience of gravity as well as the explanation for all sorts of cosmological events from black holes to Big Bangs.

Energy and matter
But there is no reason why the influence of spacetime should be restricted only to objects made of matter. Energy too can influence and be influenced by spacetime. It is this which lies at the root of the discovery that energy and matter are more similar than we tend to imagine. That matter is 'just' densely-packed energy, and that $E=mc^2$.

RELATIVITY: FAR-REACHING BUT LITTLE UNDERSTOOD
Perhaps because these ideas are so complex, so challenging, googling 'relativity' does not yield a satisfying definition; unlike say googling 'evolution'. The definition below is one of the simplest you will find.

Google

relativity
The dependence of various physical phenomena on relative motion of the observer and the observed objects, especially regarding the nature and behaviour of light, space, time and gravity.

evolution
The process by which different kinds of living organism are believed to have developed from earlier forms during the history of the earth.

Relativity's profound implications are at least as important as the Theory of Evolution to humankind's sense of the world and our place in it. Despite this it remains little understood. Worse still, beyond this initial definition of relativity the reader will meet a steep dive into physics, obscuring its real significance from the average person.

Trying to understand relativity can be dauntingly hard. But it need not be. It can be done with straightforward logic. For a proven theory with far-reaching implications it is important for an understanding of relativity to be more accessible.

AN EVERYDAY DEFINITION OF RELATIVITY

This is how I would define relativity.

> **relativity**
>
> A proven theory that describes aspects of the universe's design. It also explains that what we observe depends on how much faster we are travelling than the thing we are observing. Relativity reveals a universe radically different from the one we perceive with our everyday senses.

Asked for more detail, I would make the following statements.

- The universe's speed limit is one of the universe's fixed constants. It is the speed at which light and other electromagnetic waves travel.
- A metre of space and a second of time aren't fixed properties: space appears to contract and time appears to slow when observers move at different speeds.
- Despite our perception that space and time are separate they are in fact two different perspectives of 'spacetime'.
- Spacetime is curved by matter or energy. What we experience as 'gravity' is the result of objects following these curves in spacetime.
- Matter is made up of very high density energy.
- Relativity is one of the most important advances in knowledge there has ever been, making possible much new science and engineering.
- Relativity challenges our ideas of reality, shining light on questions about the nature of time and of the universe and our place in it.
- Relativity has sparked imaginations, influencing culture and society in diverse areas.

UNRAVELLING RELATIVITY

This book aims to unravel the mystery of relativity in the most straightforward way possible. Before we start we'll highlight a few things which will help.

1 Historical context

Awareness of the historical context from which it emerged can help us understand relativity. Physicists have been trying for centuries to understand the workings of the universe. Generally that search progresses one small step at a time. Occasionally there is a major breakthrough, such as relativity, but it is not wholly disconnected from what went before.

2 Logic, pure and simple

It is possible to appreciate Einstein's relativity with little or no mathematics and without, for some of us, its suffocating rigour.

There are those who say it isn't possible to appreciate the meaning and beauty of modern physics without some use of mathematics. This seems all the more ironic given Einstein's own use of logic. It was from his so-called 'thought experiments', simple and some even fun, that he developed his initial ideas – which he *then* underpinned with mathematics.

Einstein wasn't *impossibly* clever
Mere mortals find it hard to grasp
The quite surprising thought
Of Space and Time Equivalence
Across universe so vast.

And yet the maths it so decrees
The science proves it true
So we assume its inventor's
Brain, As large as planets be.

But for us who are not built the same
How to grasp what he declared?
To peer behind his hair and maths
With a logic more mundane?

In similar vein we can appreciate a cathedral's beauty without knowing the complex engineering maths behind the elegant but seemingly implausible columns supporting equally implausible vast vaulted ceilings. A simple explanation enhances our marvelling at the graceful power.

Of course with a deeper understanding of the architect's calculations comes an appreciation of the art of the architect and what he or she has achieved. But it is not a prerequisite to any appreciation at all. When describing the cathedral to friends we do not roll out its blueprints. We use words and pictures to describe the vaulted ceiling's scale and beauty. If asked, we'll describe how an arch can support far more weight than a flat ceiling but draw short of the mathematics which explains that technology.

3 Simple concepts about travelling from A to B
There are a few simple concepts which we'll need in this book.

How quickly I can sail from A to B depends on a several factors. These include my speed and direction of travel and whether I encounter assistance or resistance.

The greater my speed the more quickly I travel the distance. Speed is measured in 'kilometres per hour' (kph) or similar. If in 1 hour I sail 4 kilometres my speed is 4 kph. If I double my speed to 8 kph I'll cover the 4 kilometres in half the time, that is in 30 mins. In other words:

> **SPEED equals DISTANCE divided by TIME**
> or, speed = $\dfrac{\text{distance}}{\text{time}}$

(In this book, the words 'distance', 'length' and 'space' are used interchangeably to mean similar things depending on context.)

Direction matters

If I take a zig-zag course from A to B it will take more time than a straight line course. In other words direction of travel is as important a consideration as speed of travel.

Friction-free travel

Throughout this book in all the examples and explanations a simple assumption is made. When we talk about objects or light waves travelling anywhere, we assume they are doing so in a vacuum. They aren't stopped or slowed down by anything unless stated. That's just the normal convention used by physicists (except of course when they're actually trying to understand how air resistance, friction, etc. affects motion).

4 Mathematical descriptions and possible interpretations

Physicists usually try to develop a picture or working model of how the universe works and then develop the maths to support it. For instance, observations that the planets orbit the sun preceded Sir Isaac Newton's equations which explained the influence of gravity.

Sometimes it's not possible to develop a working model first, so physicists develop a mathematical description of observations and use that description to help them develop and test theories, or 'interpretations', of how the universe works.

Some of these interpretations challenge our ability to understand or believe because they don't match our brain's ideas of how the world around us works. At such times it's worth bearing in mind that despite Newton's wonderful equations which explained the influence of gravity he didn't actually know what gravity was. He had a few ideas but it could just as well have been giant invisible monkeys pushing the planets around for all he knew. Despite this we all freely climb into cars and aircraft whose engineering (not to mention our lives) depend on Newton's equations.

5 Challenging preconceptions: seeing should not limit believing

Our perceptions of the world are what most inhibit understanding relativity. We'll discover that if we travel very fast time appears to slow and length contract. Every fibre of our being rejects this obvious nonsense. That's how we humans have evolved, to subconsciously make sense of our surroundings based on our everyday experience. But we don't have any experience of travelling at vast speed, one *million* times faster than an aircraft. So we need to distrust our intuitions if they are based solely on our unconscious presumptions.

Imagine you were born and raised on a very large plain. Gently rolling fields of tall grass and trees swaying in the breeze are all that you can see. One day a stranger visits with tales of mountains taller than the sky and oceans of water bigger than the plain

you're standing on. You're incredulous, maybe even you disbelieve the stranger. But what right do you have to be so sceptical just because all that you've ever seen is a gently rolling plain?

> The highest form of ignorance is when you reject something you don't know anything about.
> *Wayne Dyer, Self-help author*

It is more likely, certainly more sensible, that you ask about this 'wild' claim. If you ask, "How can it be so?" you are asking for a *theory*. The stranger may suggest that you imagine a hill of earth and rock piled onto another hill of earth and rock, or a hole getting bigger and bigger being filled with more and more water. You may be partially satisfied by this, but remain unpersuaded. If you then ask, "Can you prove it?" you are asking for *evidence* or *verification*. At this point the stranger may show you pictures or invite you to go see for yourself. You can be assured that Einstein's relativity offers both a robust theory and plenty of evidence.

Don't Be Put Off

Logic is all that's needed to explain relativity, however understanding it is not all plain sailing. Relativity encompasses very many inter-related and challenging concepts. Don't be put off. It may well not all fall into place in one go. Keep going – you will almost certainly gain a better understanding than before you started.

I hope you enjoy the journey.

> [Relativity] appealed to me like a great work of art ... the greatest feat of human thinking about nature, the most amazing combination of philosophical penetration, physical intuition, and mathematical skill [even though] its connections with experience were slender.
>
> *Max Born, Physicist*

What to Read

The explanations of relativity are in three progressively more detailed chapters.

- *Relativity Over a Cup of Coffee.* This may satisfy your needs or it may serve to ease you into the next chapter.
- *More Detailed Logic.* If you want deeper explanations. You can jump straight to this chapter if you wish but it may well prove useful to read the *Cup of Coffee* overview first.
- *A Little Bit of Maths.* Purely optional maths if interested.

In addition there is a discussion about quantum mechanics and how it and relativity contribute to the on-going search for a 'Theory of Everything'. And there are less technical chapters: physical evidence that supports relativity theory; relativity's implications for our understanding of time and other philosophical questions; its influence on culture and society; and a chapter on Einstein himself.

Chapter
RELATIVITY OVER A CUP OF COFFEE

This chapter explains relativity in three parts.
1 Our everyday ideas of space and time are flawed.
2 A new understanding of space and time addresses these flaws.
3 This leads to further insights into how the universe works.

1 EVERYDAY IDEAS OF SPACE AND TIME ARE FLAWED

THE SPEED OF LIGHT IS REALLY VERY FAST
Travelling at over 300 metres per second sound takes one third of a second to travel 100 metres. Our brains might just notice the lag between a shout and the response of someone at that distance.

On the other hand light beams travel at 300 *million* metres per second, so called **light-speed**. So it takes a beam of light only one third of a *millionth* of a second to travel 100 metres. Our brains just cannot notice the time lag over this or indeed much greater distances.

As a result we humans believe that anything we see now is happening now. But the moon we see is actually the moon as it was one second ago, and the sun as it was 8.5 minutes ago, i.e. the time it took their light to reach earth.

THE SPEED OF LIGHT AND OUR PERCEPTION OF TIME
At age 16 Einstein asked himself how things would look if he could ride on a beam of light. Within 10 years his answer and its implications had begun a revolution in physics and transformed our ideas of space and time. His initial insight can be described as follows.

> That's it! That's the Theory of Relativity! Light travels to us from the face of a clock, to tell us the time. But, if we travel away from the clock at the speed of light the face of the clock appears to have stopped! Time would stand still. This moment would last forever.
> *Fictional dialogue between Albert Einstein and Marie Curie*
> *Adapted from the movie: Young Einstein*

Let's explore Einstein's question with an imaginary experiment.

All aboard a journey in time

A clock, like the one at position 'A' in the picture, sends out regular time signals. To the right of 'A' we see these signals as light images of the clock face which 'left' the clock earlier, at quarter hour intervals.

Imagine you're in a rocket travelling at the speed of light. At 12 noon you passed the clock at 'A'. An hour later you have arrived at position 'C' and look out of your window at the clock. Surprisingly it appears to read 12 noon. But that's because you're actually seeing the 12 noon image that's been travelling at the speed of light beside your rocket all along. You can't actually see how the clock looks *now*, only how it looked one hour ago. **Time outside your rocket appears to you to have stopped.**

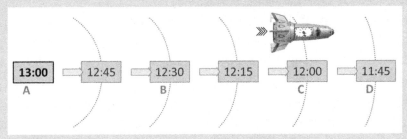

However, despite this, as you look at your watch you would believe that **time inside your rocket is ticking away normally**.

If you'd travelled slower than light-speed so that after an hour you'd only reached position 'B', the clock image outside your window would read 12.30. So, the time you observe depends on your speed relative to the clock. **The faster you travel the more time outside appears to slow.**

Meanwhile, your friend has remained at 'A'. She's certain that time has advanced normally and that it's now 13.00. It seems that **the perception of time and the concept of 'now' are local to individuals.**

If you could go faster than the speed of light, someone at position 'D', would see you arrive at 11.45, 15 minutes before you left the clock! But we don't normally see such reversals of cause and effect. Perhaps **travel faster than the speed of light is impossible.**

Wondering "What is the right time?" you look for the universe's master clock, and realise that if such a thing existed this confusion might have been avoided in the first place. (There's nothing special about the clock being at 'A': it could be anywhere in the universe with the same result. Neither would having your own watch help you know the 'right time'.) **There is no universal master clock, no single reference point of time.**

14

You might wonder if the experiment's conclusions are as imaginary as the experiment. But many real experiments say otherwise. We only believe that time has a universe-wide constant rhythm because our everyday low speed experience doesn't confront us with evidence to the contrary. Einstein's brilliance was to look beyond the everyday.

And it's not just clocks that slow. Successive images of your friend who stayed at 'A' will be more and more delayed arriving at your fast-travelling rocket: so they appear to be aging more slowly than you.

It's All Relative

If there is no single reference point for time, the start point of your journey, point A, could have been anywhere in space with the same outcomes. This suggests there is no single reference point for space either, **no privileged place from which to observe the universe**.

It follows that **position, time and speed are only meaningful when measured *relative* to other things**, in this case the clock at 'A'. It is meaningless to ask if your rocket is moving or stationary unless you also add 'by comparison to' another object.

Light Always Moves at Light-Speed Faster Than Observers

Light beams are waves of energy that have no mass, so they can travel as fast as the universe allows, i.e. at light-speed. 'By comparison' to your rocket, therefore, you must observe the light beam travelling at light-speed faster than your rocket – *no matter how fast it's moving*. It isn't what we expect, but it's also been demonstrated by experiment.

Space Appears to Contract as Speed Increases

On page 10 we saw how to calculate speed. So for a light beam:
   the speed of light equals *distance travelled* divided by *time taken*
But, as we've just seen, light beams always appear to travel at the same light-speed. So as your rocket speeds up and time slows, the 'time taken' reduces. Therefore 'distance travelled' must also reduce for the equation to give the same result of 'light-speed', no matter your rocket's speed. In other words, the faster you travel the more distance outside your rocket appears to you to contract. Meanwhile, distances inside your rocket will appear to you unchanged.

2 A NEW UNDERSTANDING OF SPACE AND TIME

WHAT'S ACTUALLY GOING ON WITH SPACE AND TIME?

Do time and space really change or just appear to do so? Experience leads us to expect: that a one metre ruler is always one metre long; that one second of time is always one second long; that time and space are unrelated; and, that this is the case everywhere in the universe. But, maybe our low speed intuitions have not prepared us for life in the fast lane, at speeds approaching that of light. How can we know it's true at speeds one million times faster than our experience? After all in real experiments space appears to contract and time appears to slow. Just what is going on?

It's worth pausing for a subtle but important definition. '**Appears to**' could mean: "looks like, but isn't really", but here it means: "this is how it actually is to *for that specific individual observer*".

We've seen that the faster we travel the more space appears to contract *and* the more time appears to slow. In other words, the length of one metre decreases *and* the duration of one second increases. This hints at a relationship between space and time, a relationship in which they trade off, one against the other.

By way of an analogy imagine a long thin box. Held horizontally it can't pass through a doorway. However, rotated to the near vertical it passes through easily. We might say that the box's width and height have traded off against each other.

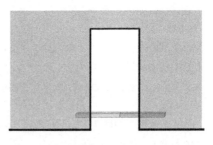

Imagine now an alien unfamiliar with fanciful ideas such as up & down. He has only a two-dimensional (2D) view of reality: left & right, forward & backward. He can't see the rotation into the vertical. He only sees that the box appears narrower and passes easily through the doorway. We humans, aware of three dimensions (3D), understand that the key to what the alien perceives as 'magic' is the rotation of the box.

Something similar happens with space and time. Only now, it's we 3D-limited humans who are unaware that space and time are both parts of four-dimensional '**spacetime**', in which the length of one metre *and* the duration of one second trade off against each other. Physicists don't know what spacetime looks like, but that doesn't mean the concept is flawed. (Recall that Newton didn't know what gravity was but his maths worked well.)

Put another way: the faster we travel through space, the more slowly we travel through time, and vice versa. This is not as odd as it sounds. We often say that time flows, meaning 'time flows past us'. But we could visualise it as 'us' moving through 'time'. And, if we can move in time as well as in space, this trade-off is easier to imagine.

Back to our imaginary experiment from page 14. As your friend by the clock at 'A' looks out into space she's actually looking into spacetime. She watches you fly past in your rocket, and sees that one second on the rocket's clock is longer than one second on hers, and that as a trade-off, one metre of rocket length is shorter than her metre ruler. This trade-off increases as relative speed increases. It is the key to the 'magic' of relativity. (This is different to things looking smaller when they're far away. This trade off of relativity would be noticed even if the rocket flashed past just metres away from the clock.)

If space and time are affected by the motion of objects, then spacetime must be as much a part of the universe as the objects themselves. This is different to our intuition that space is a fixed stage on which the universe plays out its story to the constant beat of time.

One further point. In the picture on page 16 the left and right ends of the box are differently coloured. If the box could rotate beyond the vertical, left and right would swap round. Such a reversal of the alien's reality would be hard to accept, and so maybe the box cannot (for some reason) rotate beyond the 'limit'. The universe also has a limit – a **speed limit**. As we saw in the experiment on page 14, travelling faster than light-speed meant that you'd arrive at position 'D' before someone at 'D' received the images of you leaving the clock at 'A'.

Is All this Real?

It's tempting to regard the trade-off between space and time as a sort of illusion; no more magical than the box and doorway. After all once the box is horizontal again the 2D-thinking alien will think all is back to normal. Similarly, once relative speed returns to zero the lengths of rulers and the rates of clock on the two rockets will once again agree.

But while the box has returned to its normal width the alien nonetheless finds the (now delivered) box on the other side of the doorway. The result of the rotation is part of the alien's reality even though the rotation itself is beyond his comprehension.

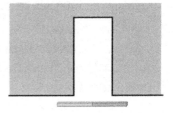

Likewise, length contracting and time slowing become part of our reality. For example, satellite navigation systems are corrected for time slowing. Without this, actual and sat-nav position would differ.

Furthermore, we know that in 3D the box didn't actually change shape. Equally, in 4D spacetime the combined 'interval' in both space and time stays constant despite space contracting and time slowing.

And Vice Versa – It's All Relative

We saw on page 15 that speed can only be described relative to another object. This means that the question, "Is that rocket moving forwards at constant speed past me or am I moving backwards at constant speed past it?" does not have a right answer.

So, while your friend by the clock sees time slowing and length contracting in your rocket, while everything appears as normal by the clock; from your rocket you will observe the exact opposite: that time has slowed for her and she and the clock have contracted. In other words, you each observe the universe from your own 'bubble of reality'. Neither of you are privileged in any way. This is how things actually *appear to* each of you as *specific individual observers*.

At this point there might seem to be a problem. How can the other person's length appear shorter and its clock slower to each of you at the same time? However, this isn't the paradox it appears to be as is explained by the *Twin Paradox* on page 40.

3 NEW INSIGHTS INTO HOW THE UNIVERSE WORKS

A QUICK SUMMARY

Our imaginary rocket has taken us on quite a journey. Its surprising conclusions are, however, supported by experiment as well as the maths of relativity. Let's summarise what we've found so far.

- Speed is meaningful only by comparison to other objects.
- As relative speed between objects increases, observers at one object will see another object's time slow and length contract.
- At low relative speeds these changes are too tiny to notice.
- When relative speed reaches light-speed, time will appear to have stopped and length will appear to have contracted to nothing.
- Nothing travels faster than light-speed, the universe's speed limit.
- Only massless waves such as light can travel at light-speed; being without mass they travel at the fastest speed the universe allows.
- Space and time are not absolute, unchanging entities. They are interrelated, as one appears to change so does the other.

MASS INCREASES TOO

Imagine a rocket fires its engine non-stop. We might expect it to keep gaining speed, eventually exceeding light-speed. But as we've seen that's not allowed. So where does all that extra engine energy go?

A slow-moving heavy lorry will damage a brick wall just as much as a fast-moving car. We see from this that the energy that causes the damage arises from the vehicles' speed as well as their weight. So, with our rocket, at low speeds the extra engine energy results in extra speed. But when already moving fast, it results in extra weight. Of course the rocket has not gained bulk, but it behaves *as if* it is heavier.

This is not as strange as it sounds: a rocket weighs more on earth than when weightless in space. Its **weight** depends on *where* it is. Relativity tells us it also depends on its *speed* relative to an observer.

Once again, this may seem like an illusion, but yet again it has real world consequences. For example, particle beam accelerators, as used in radiotherapy, need increasingly powerful magnets to deflect the increasingly heavy particle beams as they approach light-speed.

Mass and Energy, Why Does E=mc²?

We've just seen that at vast speeds if an object tries to go even faster, the extra engine energy results in extra weight rather than extra speed. Since an object's weight is related to its mass, this hints at a relationship between energy 'E' and mass 'm'. Einstein imagined a light beam colliding with an object and transferring its energy to the object. By combining the equation for the energy of a light beam with the equation for the additional motion energy gained by the object, he found that relationship to be $E=mc^2$, where 'c' is the speed of light.

Since none of the three terms in the equation refer specifically to collisions or rockets, Einstein realised that $E=mc^2$ implied a more general **equivalence between mass and energy**. This idea confirmed what physicists working in another area had already noticed: when radioactive atoms decay they lose mass and they also radiate energy.

Since the speed of light 'c' is a big number, so c^2 is huge: 10 followed by sixteen zeros. As such, the equation $E=mc^2$ tells us that the energy locked up in even a tiny amount of matter is vast. From this we can infer that maybe matter is 'just' very densely packed energy. It is this that lies behind the opportunity and threat of nuclear reactions.

A further important idea lies behind $E=mc^2$. Everything we see in the universe, whether matter or tiny particles, is made of energy. The universe is not *matter-based* as we imagine but *energy-based*. It is the interaction between different densities and arrangements of energy that gives rise to all that we experience and all that we see.

How much energy is there in a rugby player?
If all the matter in an average-sized 100kg male rugby player could be converted to energy it would meet the UK's annual energy demand.* If instead the unfortunate rugby player was burnt on a fire he'd barely warm a family for an evening. (NB Nuclear reactors actually convert only about 1% of mass into energy. To understand why, see page 51.)

* See *A Little Bit of Maths*, page 122, for the calculation.

It's Gravity – But Not As We Know It

We've seen that when we travel at light-like speeds length, time and mass appear to change in ways that have real world consequences. Since these quantities underpin the laws of physics the laws needed to be revised. This included Sir Isaac Newton's laws of 1687 describing the force of gravity. It had worked well for 200 years, despite Newton having no idea what actually gave rise to gravity.

Einstein realised that our experience of gravity is the same as being in a rocket that is increasing in speed. The pressure felt between our feet and the floor is the same, just as we feel heavier in a lift as it starts to rise. Gravity, he realised, is not a force pulling objects towards each other, as we imagine, it arises from a pressure on objects. But if that's the case what is that pressure, where does it come from?

As you read the following, you may find it helpful to keep in mind the analogy of a metal object in the presence of a magnet: it feels a pressure to move.

On page 17 we saw that spacetime is part of the universe. Einstein went on to explain that objects interact with **spacetime causing it to curve**. We can visualise this effect as a curved grid, as below.

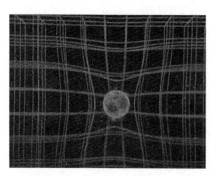

Since spacetime describes the landscape which contains all objects, other objects will feel a pressure to follow these curves, much like a ball rolling down a hill. At times they 'fall' in toward the centre of these 'gravity wells', at other times they might be held in an orbit around them. What we think of as gravity 'pulling' on objects is actually the pressure of spacetime's curves directing their motion.

We don't know what spacetime consists of, nor how the densely-packed energy that makes up objects interacts with it. But this theory of gravity has been proven. See page 67, *Where's the Proof?*

THE SPEED OF LIGHT 'c' IS REALLY VERY SPECIAL

We've seen that light-speed 'c' plays several key roles. It is the universe's speed limit, it defines the relationship between space and time (spacetime), and that between mass and energy ($E=mc^2$).

It is as if we are peering deep into the universe's design. The universe is not defined by rigid space and absolute time, as we perceive, but by a constant 'c'. It is important to understand here that it is not light that is special, but c. It is this universal constant which limits light to the 'light-speed' we observe.

IT'S NOT ALL RELATIVE

The laws of physics utilise fundamental properties such as distance, time and mass in equations to describe how our universe works. These *laws* are the same throughout the universe. However, the *values* we observe for these properties vary, depending on the *relative* speed between objects. Hence the name 'relativity'.

HOW TO APPEAR SLIM AND STAY YOUNG LOOKING

If you're able to travel at light-like speeds you'll appear slimmer and younger to your friends – but only if they aren't travelling with you. (If they were, there would be no relative speed difference, so it wouldn't be much of a 'comparative' experience!) Not only that but, because motion is relative, when you look at your friends, you will think they look in better health than you!

CONGRATULATIONS!

Understanding relativity can be challenging. Hopefully you've had the opportunity to appreciate the beauty of relativity. By way of a summary you may wish to return to the definition of relativity on page 7. And remember ...

> The universe is under no obligation to make sense to you.
> *Neil deGrass, Astrophysicist*

NOW THAT YOU'VE FINISHED YOUR COFFEE

CHALLENGING PRECONCEPTIONS: REALITY IS NOT WHAT IT SEEMS

It comes as a surprise to discover that our age-old intuitions of space
and time derive solely from evolution-constrained senses, senses
born in a low speed world. It transpires that these are an unreliable
guide to how the universe beyond our everyday senses actually
works. Our observed reality is just a universe-sized illusion.

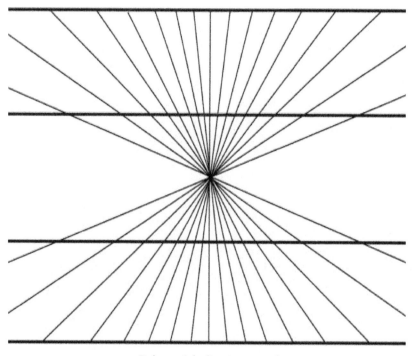

Only straight lines are used
in this illusion of curved lines.

Relativity has been proven, see the chapter *Where's the Proof?* on
page 67, which includes examples of how relativity theory enables the
building of practical machines. That's the ultimate test of any theory,
whether it works or not. In its day-to-day work, Science is not a quest
for truth but for ever-evolving theories which explain with increasing
accuracy how the world works.

The logic of this chapter is explored more deeply in the next, *More Detailed Logic*. But if this explanation has been sufficient for you then skip that and go on to read how relativity has informed or influenced the on-going search for a Theory of Everything, our understanding of time and other philosophical questions, and how it has influenced culture and society.

Chapter

MORE DETAILED LOGIC

This chapter explores more deeply the logic summarised in the previous *Cup of Coffee* length description of relativity. Although that chapter is not required reading before this one, you may find it helpful to have gained that overview first.

This chapter is set out as follows.

1   Before relativity
   'classical' physics had been working so well.

2   Simple yet profound questions
   reveal flaws in the classical theories.

3   Time slows, space contracts and mass increases
   as speed increases – strange but true.

4   A radically new view of the universe:
   goodbye space and time, hello spacetime.

5   Revising the laws of physics
   reveals new perspectives on matter, energy and gravity.

6   Insights into the universe's design:
   an expanding universe, defined by the speed of light.

7   And finally,
   a few loose ends.

A Few More Concepts About Travelling From A to B

Before we continue however, we need to add a few concepts to those on page 10. Don't worry if you don't understand them fully now.

Changing speed or direction. Mass, matter and weight
Imagine you're travelling in a ship. If you change speed or direction it affects how quickly you travel from A to B. Physicists call any change to speed or direction or both '**acceleration**'. (NB In everyday speech the word 'acceleration' is incorrectly used only to changes in speed.) A ship that is not accelerating is described as moving '**uniformly**'.

To change speed or direction requires '**force**', such as the wind, to overcome the ship's inclination to carry on doing whatever it's already doing. The amount of this resistance, known as '**inertia**', is quantified in terms of kilograms of '**mass**'. It depends on how much material, or '**matter**', the ship is made of. It is because of this connection that we tend in everyday speech to wrongly equate 'mass' and 'matter'.

The more matter the ship has, the greater its inertia, and hence mass, will be. It will require greater force to its change speed or direction, which in turn makes the ship appear heavier. When something is 'heavier' we say it has more '**weight**'. However, the experience of weight is also a function of *where* an object is. In space, where there is minimal gravity, an object is weight-*less*.

There is one more thing: understanding acceleration and force in the special case of rotating objects.

Rotating objects
Consider an object on a spinning sphere, for example, *you* standing on earth. You feel you're not moving because your speed is zero *relative* to the earth. But, as you travel in a curved path, you're changing direction, and so *accelerating*, according to the definition of acceleration above. The definition also says that for an object to accelerate there must be a force. In this case, the 'force' is provided by gravity, preventing you from being thrown off the spinning earth and travelling out into space in a straight line.

1 BEFORE RELATIVITY

GALILEO, NEWTON AND MAXWELL

In 1632 Galileo Galilei considered a sailor in a windowless cabin of a ship on a calm sea. He said the sailor would be unable to tell whether his ship was stationary or moving 'uniformly'. Even if the

sailor had a window from which he sees another ship appearing to sail past uniformly he'd have no way of knowing which ship, his or the other or both, was actually moving. Galileo's sailor can only describe his position or motion *relative* to the other ship and not to any universe-wide reference point. This is known as the '**Principle of Relativity**'. (A better term for '**stationary**' would be '**co-moving**' meaning 'not moving relative to each other'. But that's cumbersome to use, so this book uses the term 'stationary' in inverted commas.)

Furthermore, on this ship an apple thrown straight up would fall straight down into the hand which threw it, just as when you throw an apple upwards on a moving train. If it didn't, Galileo's sailor would know if his ship is moving or not and that would violate the Principle of Relativity. Only if the ship accelerates, that is changes speed or direction, would the apple fall away from the sailor's hand.

Later, in 1687, Sir Isaac Newton explained how the universe worked. His theory combined motion on earth and in space into a model which matched observations and predicted events. This brilliant model accepted the Principle of Relativity. It also assumed that space is like a rigid container and that time has an absolutely constant rhythm, marching to the beat of the same clock everywhere.

> ### Cracks in physics
> Newton's theory of gravity worked near perfectly (albeit, there was a tiny inaccuracy when calculating the orbit of the planet Mercury). But, despite the success of his theory, it didn't explain what gravity actually was. Nor did Newton know if the universe actually had a one and only constant rhythm, a 'true time', even though his theory assumed it did.

In 1861 James Clerk Maxwell explained that electricity and magnetism weren't separate things, as we perceive them, but aspects of the same phenomenon. We encounter this phenomenon daily in the form of radio waves, or **electromagnetic waves**, such as used by mobile phones and wi-fi. Maxwell's maths gave the speed of these waves in a vacuum as nearly 300 million metres per second. This was exactly the same as the speed of light which had been measured with increasing accuracy since the 1600s. It seemed certain that light must be an electromagnetic wave. Einstein said of Maxwell's work that,

'This change in the conception of reality is the most profound and the most fruitful … since the time of Newton.'

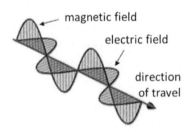

magnetic field

electric field

direction of travel

Light is an example of an electromagnetic wave. These waves are self-propagating: the vibrating electric field stimulates a magnetic field (at right angles to it) and vice versa. And, unless obstructed in some way, there is nothing to interfere with this continued action.

The fact that Maxwell had shown that light was a wave, an electromagnetic wave, was good news. Physicists understood waves such as ripples in water or sound in air. The question they turned their attention to now was, light in 'what'. And so the search for the 'what' started; a search for a concept first suggested by physicists in 1690, and which had been called the **'ether'**.

More cracks in physics

In 1887, **Albert Michelson and Edward Morley's** search for evidence of the ether surprised physicists. In an experiment a light beam was split in two and the two halves sent off at right angles to each other, reflected back, and then compared in a detector to see if their wave peaks and troughs still coincided. They expected that the speed of the beam which moved in the same direction as the earth's rotation would be affected by that rotation, while the beam at right angles to it would remain unaffected. But no difference between the two beams was observed. They concluded that light will always appear to travel at light-speed *faster than an observer*, no matter how fast the light source or observer is moving. This is known as the **'invariance of light-speed'**.

ALWAYS THE SAME, INVARIANT LIGHT-SPEED

Michelson and Morley's 'failed' experiment led to the physicists' nightmare described on page 13. In reality it was no more than what Maxwell's electromagnetic wave theory had already made clear: the speed of electromagnetic waves depends *only* on the magnetic and electrical properties of the medium through which they travel[1]. In other words, the speed of electromagnetic waves *did not* depend on anything else, such as the motion of light sources or observers.

This is the main reason that the mathematical symbol for light-speed is **'c', for constant**. (Another reason sometimes given is that 'c' also stands for 'celeritas', Latin for 'swiftness', a term fully justified given that light travels at one million times our everyday experience.)

PHYSICS, JOB (NEARLY) DONE? ERR, NO

Newton's and Maxwell's brilliant theories unified features of the world we observe. While on the face of it the theories appeared to have little to do with each other physicists wondered if further unification was possible. If only they could find the light-carrying ether maybe it would also turn out to be the fabric of Newton's rigid universe.

Newton, 1687
Gravity:
Motion in space
Motion on earth

Maxwell, 1861
Electricity
Magnetism
Light

Despite the so far failed search for the ether and the other cracks in physics a few physicists felt optimistic in the late 1800s. Having explained all then known forces and types of energy[2] they wondered if all that was left was some fine tuning before declaring 'job done'.[3]

However, the cracks in physics didn't yield to fine tuning. Einstein himself spent seven fruitless years on them before concluding that the old physics was broken, as we shall see in the next few pages.

---

[1] Even what we imagine as empty space has these properties.

[2] Including heat, see Thermodynamics on page 89.

[3] Not only had they not anticipated relativity, but also the atom-splitting of 1917 revealing the atom's inner structure and new nuclear forces. (These are described in the chapter *Theory of Everything, Quantum Matters*, page 71.)

2 SIMPLE YET PROFOUND QUESTIONS

ENTER EINSTEIN, TRAVELLING ON A BEAM OF LIGHT

Einstein first started having doubts about 'space' and 'time' in 1895, aged just 16. He asked himself how the world would look if he could travel on a beam of light. He couldn't have known that this simple question would unlock a completely new view of the universe; or that it would unify existing theories, putting him on a par with Newton and Maxwell.

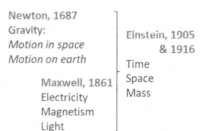

Newton, 1687
Gravity:
Motion in space
Motion on earth

Maxwell, 1861
Electricity
Magnetism
Light

Einstein, 1905 & 1916
Time
Space
Mass

Thought Experiment #1: Travelling at the speed of light
In the early 1900s Einstein was still wondering how the world would look if he could travel on a beam of light. Each day he boarded a tram to go to work at the Swiss Patent Office in Bern. He imagined his tram travelling at the speed of light[*1] away from a clock tower, starting its journey at noon.[*2]

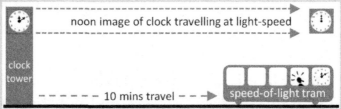

noon image of clock travelling at light-speed

clock tower

— — — — — 10 mins travel — — → speed-of-light tram

Einstein realised that as he sat on the tram looking out of the window the tower's clock would appear frozen in time at 12.00 noon, because both he and the light of the noon image are travelling at exactly the same light-speed. He also realised that the tram's clock would be ticking as normal, so after 10 minutes it would show 12.10!

Importantly, the result would be the same irrespective of starting point anywhere in the universe and irrespective of direction of travel; again undermining the idea of absolute reference points for space.

[*1] Of course we now know that nothing, other than things like light, can travel at light-speed. But before relativity the old view of physics ruled. It was the presumptions of the old physics that Einstein was putting to the test here.

[*2] Despite this story about Einstein being widely and reputably used, there is some evidence it is a myth. Nonetheless, no one doubts its scientific value.

WHAT'S THE RIGHT TIME? THE RELATIVITY OF SIMULTANEITY

Thought experiment #1 hints at a relationship between the tram's speed and time itself. Such a relationship undermines the idea of absolute time, an idea about to receive a further fatal blow.

How can tram-riding Einstein know the actual time? Is there even a universe-wide 'right time'? If we ask ourselves, "What is time?" we might answer, "Time describes the gap between two events". If there is such a thing as absolute time, Einstein reasoned, all observers should agree on the gap in time between two events. He then went on to show that that's impossible using the thought experiment below, thus putting an end to Newton's assumption of absolute time.

Thought Experiment #2: No such thing as 'the right time'
A moving train flashes lights at both ends of its carriage. A bystander sees the two flashes simultaneously because they travel the same distance to his eyes. However, a passenger on the train will have moved forwards with the train by the time the flashes reach her. And, since she's moving away from one source and towards the other, the light flashes' journeys are different and won't appear simultaneous to her.

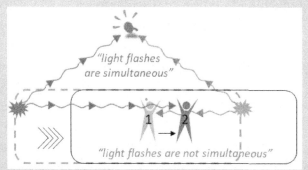

We conclude that observers in motion relative to each other can't agree on whether two events are simultaneous or not. The bystander and passenger will even disagree on terminology: for example "light flash arrives *now*," will cause confusion about what '**now**' means.
NB We might imagine that since both observers in this experiment understand that light-speed is the same for all observers no matter how they move, they should be able to work out their disagreement. But to do that they would need to (a) synchronise their clocks, so as to (b) communicate to each other when the light flashes arrive. However, since such synchronising signals are themselves constrained by light-speed, Einstein could demonstrate this wasn't possible.

Summarising the Problems with the Idea of Absolute Time

- The Relativity of Simultaneity shows that two observers can't agree whether two events are simultaneous. (In the same way, in space, asking if two objects are at the same height is only meaningful by reference to, for example, the surface of a planet.)

- In his light-speed tram Einstein thinks the clock tower's clock has stopped. And, if the tram travelled at less than light-speed, time would appear to slow (as the rockets on page 15 found). If there is no clock on the tram, Einstein's only knowledge of time is the clock tower's seemingly unreliable image.

- Light-speed is so fast that light takes just $1/1000^{th}$ of a second to travel 250 kilometres, so we tend to imagine that anything we see happening is happening right '**now**'. But since light-speed is finite, we actually see things as they were when the images we do see 'now' left those things a while ago. It turns out that 'now' is not a well-defined concept. There is no universally agreed 'now'.

- This isn't only about clocks. Just as the clock tower's image is delayed reaching the tram so will all other images. For example, living organisms will appear to age more slowly. Light also conveys **information**, for instance the time at the clock tower. So, the speed of information transfer is also limited by light-speed.

Challenging Preconceptions: Everyday Speed v. Light-speed

None of this should come as such a huge surprise. We don't actually know what happens at speeds *one million* times that of our everyday experience. It seems sensible not to presume. For example, we wouldn't take just one millionth of the Bible's 750,000 words and pretend to understand Judaism and Christianity.

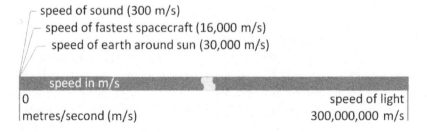

- speed of sound (300 m/s)
- speed of fastest spacecraft (16,000 m/s)
- speed of earth around sun (30,000 m/s)

speed in m/s

0
metres/second (m/s)

speed of light
300,000,000 m/s

THE OLD PHYSICS IS BROKEN, SOMETHING'S GOT TO GIVE

As mentioned previously Einstein came to realise that the old physics was broken, that either Newton or Maxwell was wrong in some way. Either time was absolute (as per Newton's equations), or time could appear to slow (as per Einstein's tram ride and Maxwell's equations).

Newton's equations and the assumption of absolute time		Maxwell's equations and the invariance of light-speed
? *Equations worked – but only nearly perfectly.* ? *No actual proof that time was absolute.*	versus	✓ *Equations worked perfectly.* ✓ *Conclusions of thought experiments.*

Maxwell's theory seemed without flaw. And, since it was in Newton's theories that the cracks in physics had appeared, Einstein concluded that something in Newton's model must be broken. A key moment in Einstein's thinking came when he realised that Maxwell's new Electromagnetic Wave Theory complied fully with Galileo's Principle of Relativity. He arrived at this conclusion through a thought experiment known as the '**magnet and conductor**'[4].

Einstein expressed the significance of this in his first paper on relativity, '*Examples of this sort, together with the unsuccessful attempts to discover any motion of the earth relatively to the [ether]*[5] *suggest ... no properties corresponding to the idea of absolute rest,*' an idea which Newton's model had assumed.

WARNING

The next 9 pages are the hardest bit of relativity: it's not that they're hard to understand, just hard to believe.

---

[4] An electrical conductor and a magnet move uniformly relative to each other. The current induced in the conductor is the same whether the conductor or the magnet is regarded as 'stationary'. It was previously thought that a conductor could be at 'absolute rest', i.e. 'stationary' in space, in which case no current would be induced.
[5] In later years Einstein made contradictory remarks about whether he had been aware of the Michelson-Morley result at the time he developed relativity theory.

THE UNIVERSE'S SPEED LIMIT

Einstein now made a conceptual leap. All the problems could be avoided if nothing other than light[6] could actually travel at or faster than light-speed. There are several reasons to suspect this might be the case.

1. If the tram that Einstein is riding could actually travel at light-speed, time would appear to stop. And if it could travel faster than light he'd catch up with images from the clock tower's past.

2. Consider two connected events: a *cause* leading to an *effect*. If some signal or object could travel faster than light we might be led to believe that the effect preceded the cause. We don't tend to see such things; for example, seeing a faster-than-light car arrive at our house before we see it turn into our road.

These first two reasons are weak: based only on the fact that we don't see such things. The next two however are stronger as they would contravene a foundational law of physics.

3. If we could travel at light-speed alongside an electromagnetic wave such as light, the Principle of Relativity implies that the light beam's waves would appear 'stationary' and therefore, from our perspective, unable to transfer its light energy from place to place. But, to an observer moving at a different relative speed, that very same wave of light would continue its illuminating journey.

4. Imagine a particle fired forwards at three-quarters of light-speed from a rocket also moving at that speed. To an observer on the rocket, the particle appears unable to overtake a light beam. But a 'stationary' sees the particle moving at its speed plus the rocket's speed, a total which is faster than light speed. If this conclusion of 'old physics' were true the two observers could tell which of them was moving, once again contravening the Principle of Relativity.

That's why Einstein described light-speed as the **universe's speed limit**. It was the starting point from which he built his new theory.

---

[6] That is, other than light and (a) all other electromagnetic waves, and (b) the information carried by these waves.

3 TIME SLOWS, SPACE CONTRACTS & 'MASS' INCREASES

THE STARTING POINT OF RELATIVITY

Einstein had concluded there was a problem with Newton's model of the universe and specifically with the assumption of absolute time upon which it was founded. This was revolutionary. It would mean creating a whole new physics since so much of the old physics was built on Newton's Laws.

Previous attempts to tweak the old physics had led nowhere, so Einstein started from a blank sheet of paper with two statements which he felt were indisputable. He called these his postulates.

First Postulate

The laws of physics operate in the same way everywhere, irrespective of relative motion.

In other words, Einstein was reasserting the Principle of Relativity, but applying it to objects moving at near light-speed as well as at everyday speeds. That's why the title of his first relativity paper was '*On the Electrodynamics of Moving Bodies*'.

Second Postulate

The speed of light (in empty space) as measured by any observer has the same value, no matter if or how those observers are moving relative to the light beam's source.

In other words, Einstein is taking the invariance of the speed of light as a given, stating that it is a universal law.

A NEW MODEL OF PHYSICS ... IN WHICH TIME SLOWS!

Having set out his two postulates, the challenge was to find a way that they might both fit into a model of physics. Let's see what happens when Einstein's two postulates are put to work together.

Let's start by recalling what Einstein's tram ride revealed: that the faster the tram travelled through space, the more slowly it appeared to travel through time, as if speed in one was trading off against speed in the other; in other words, as if they were related.

Thought Experiment #3: Time slows

A rocket astronaut and a 'stationary' observer use so-called 'light-clocks' to measure time. The astronaut sees the light beam go straight up and then be reflected straight down by a mirror, irrespective of the rocket's speed; just like an apple thrown straight up into the air on a moving train. This return journey is like the tick-tock of a clock's pendulum.

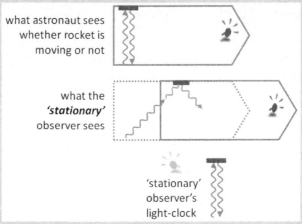

what astronaut sees whether rocket is moving or not

what the *'stationary'* observer sees

'stationary' observer's light-clock

However, since the rocket is travelling at speed to the right, the observer sees the light of the astronaut's light-clock travel forwards as well as upwards: taking a diagonal trajectory. Since light-speed is invariant (the same for everyone) the observer sees that the astronaut's light clock has not completed its longer journey by the time her clock has completed its journey. So, to the observer, astronaut time appears slower than her own. And as relative speed increases so the light beam's diagonal journey becomes longer and the slowing of time appears greater. So, from the 'stationary' observer's perspective the faster the rocket travels through space the more slowly its clock ticks.

Notice that we have applied exactly the same laws of physics, including the invariance of light-speed, to both the astronaut and observer, thereby complying with the Principle of Relativity.

We can ask if a light-clock is somehow different to a real clock. It isn't: if a real clock placed beside the moving light-clock did not slow, the two would show different times. The astronaut would then be able to know she was moving, and that would contravene the Principle of Relativity. In any case, tram-riding Einstein's problem was with real clocks, as we saw on page 30. (NB Light-clocks are used as they're a helpful starting point for the maths of relativity. See *A Little Bit of Maths*, page 113.)

EACH TO HIS OWN BUBBLE OF REALITY

The astronaut and observer each view the universe from their own **'bubble of reality'** known as a **'frame of reference'**. Each frame is an observer's individual coordinate system. It includes any object moving at low speed relative to the observer, even if they're not close by. It doesn't include any object moving at high relative speed; even if it is really close to the observer, it would be in a different reference frame. By contrast, Newton saw the universe as a single reference frame.

PROVING THAT TIME SLOWS

The slowing of time (known as **time dilation**) has been proven by experiment using highly accurate atomic clocks. (See the chapter, *Where's the Proof?* page 68, 'proof of time dilation'.)

NOT JUST TIME SLOWING, BUT LENGTH CONTRACTING

Let's place a second light-clock, at right-angles to the first, into our rocket which is moving at near light-speed. The Principle of Relativity requires that both light-clocks must run at the same rate; otherwise the astronaut would be able to tell if her rocket was moving or not.

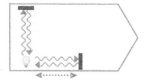 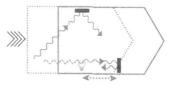

What the 'stationary' observer sees when relative speed is zero.

What the 'stationary' observer sees when relative speed is high.

At high relative speed, since the vertical clock appears slow to the observer, the horizontal clock must also tick slowly. Whilst its beam doesn't have a diagonal journey, its mirror is moving away from the light source. Since light-speed is invariant the only way both clocks can tick at the same rate is if the horizontal return journey is shorter. This means that the light-clock (and rocket, too) must appear to the observer to have contracted, though only in the direction of travel.

The observer can only conclude that, as with his idea of absolute time, Newton was also wrong about space being rigid.

This shouldn't be a surprise. Length contraction is actually a logical consequence of time slowing and the Relativity of Simultaneity. Given that 'distance' is the gap between two points measured *at the same time;* then if time appears different to two observers moving relative to each other, we should expect distance to appear different also.

PROVING THAT SPACE CONTRACTS IS NOT SO EASY

The contraction of space is known as **length contraction**. Trying to measure it directly with the equivalent of a ruler isn't possible. It would mean remotely measuring the length of an object travelling at vast speeds past 'stationary' measuring equipment. Fortunately, there are other proofs of length contraction. See the 'unexpected behaviour of muons' on page 68 in the chapter *Where's the Proof?*

BY HOW MUCH DOES LENGTH CONTRACT AND TIME SLOW?

At low speeds length contraction and slowing time are unnoticeable. But at vast speeds the amount of change is large as the graph shows.

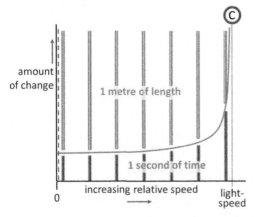

The curved line shows the amount by which length contracts and time slows as speed increases. The world of our everyday experience lies to the left of the dashed line. This is why relativity seems so strange. The curved line is called The **'Relativity Factor'**. (See *A Little Bit of Maths*, page 113, to see how it is calculated.)

For example, at the speed of sound length contracts by just 5 millionth-millionths, something we'd never notice. Alternatively to cause noticeable length contraction, for instance a 1 metre ruler contracting by 1 millimetre, the ruler would need to travel at 14 million metres per second, that's 5% of the speed of light. Nothing in our everyday experience approaches that; not even the earth's orbit around the sun comes close to that speed.

How We Experience Time Slowing and Length Contracting

 Let's return to Einstein in his tram. As he hurtles down the street, the buildings appear thinner because length has contracted in his direction of travel[7]. (Since he's not travelling vertically there is no vertical contraction.) He'd also notice the street's clocks ticking more slowly.

Let's look at why this is. Since there is no universal reference point for space or time, we each observe the world from within our own bubble of reality, our own reference frame. Within our own frame, objects move at low speeds relative to each other and classical physics works fine. But at high relative speeds things are different.

The observer in the Black reference frame at right sees a rocket (and its White reference frame) travelling at high speed  relative to him. He sees that rocket time appears to pass more slowly than his (be that ticking clocks or human ageing); and, that the rocket's length is shorter (while its height remains normal).

 But the Principle of Relativity means that the opposite must also be true, as at left. So the rocket astronaut regards his White reference frame as 'stationary' and the Black bubble, that is the rest of space, as moving backwards past his rocket. This means that as relative speed approaches light-speed, those on the rocket observe that space contracts more and more (in the direction of travel) and that any clocks in that space run more slowly.

---

[7] This characterisation of what an observer sees is helpful when learning about relativity, but it isn't fully accurate. Since light-speed is finite and because of the observer's vast speed the image is profoundly distorted in other ways too.

Hang on a Minute. Some Questions and Answers

An unsurprising question at this point is, 'Maybe objects just look shorter because they are far away?' But if that were the case they would look smaller in all dimensions (length, height and width) and not just in the direction of travel. Not only that, but length contraction is actually still observed if a rocket flashes by just metres from an observer.

Another question is, 'Is this all real or just an illusion?' The answer is 'real' and is explained on pages 16-18, *What's actually going on with space and time?* That explanation also helps us understand how two rockets travelling at different speeds can and will observe the same section of space contract by apparently different amounts.

Other teasing questions surface in these *apparent* paradoxes.

Paradoxical twins. Paradoxical ladders and barns

The **'Twin Paradox'** involves one of a pair of twins setting off in a rocket leaving the other behind, travelling a vast distance at enormous speed, and then turning around and returning to her twin. If relative speed is all that matters each twin will observe time slowing for the other twin. Each should believe the other is younger. Since that's impossible, relativity must be wrong – so the paradox suggests.

However, this *apparent* paradox presumes that the relative motion of the twins is symmetrical. In actual fact while the stay-at-home twin's course through space has remained unchanged, the travelling twin has turned around to make her return journey. This change in her frame of reference means that it is the travelling twin who ages less. (See page 111 for a fuller explanation.) Of course, if both twins had changed course at the same time and met up again somewhere in the middle, their ages would be in complete agreement.

In another **paradox, a ladder's** 'stationary' length is slightly longer than a barn. From the point of view of an observer beside the barn the ladder, travelling at vast speed, contracts and can fit inside. From the point of view of an observer travelling with the ladder the barn contracts and so the ladder cannot fit inside. However, this paradox can also be resolved. When we say 'fits inside' it implies both ends of the ladder in the barn *at the same time*. But the Relativity of Simultaneity tells us that 'at the same time' means different things to each observer. (See page 111 for a fuller explanation.)

ADDING SPEEDS – STILL CAN'T GO FASTER THAN LIGHT-SPEED

Imagine a rocket travelling at 60% of light-speed past a 'stationary' observer. It fires a missile forwards at 60% of light-speed relative to the rocket. In Newton's old physics we add the two speeds to find the speed of the missile as seen by the observer. That would equal 120% of light-speed which, impossibly, exceeds the universe's speed limit.

Instead, to find the missile's speed relative to the 'stationary' observer we must make relativistic adjustments to the missile's speed as measured by the moving rocket: that is to both the *distance* travelled and to the *time* taken. The missile's speed relative to the observer is then only 88% of c. And if, instead, the rocket and missile each travelled at 90% of c, the combined speed would be 99% of c.

A similar argument can be made for the closing speed of two rockets travelling towards each other. (See *A Little Bit of Maths*, page 116, for more detail on both these scenarios.)

AND MASS INCREASES TOO

We can extend this thinking to a single rocket. Imagine it sets off from a launch pad next to a 'stationary' observer. It fires its engine a fixed amount to get to 60% of c. It then fires its engine again by the same amount hoping to gain another 60% of c, but instead the 'stationary' observer sees the rocket's new speed as only 88% of c, as above.

Why did doubling engine-energy not double the speed? From the 'stationary' observer's perspective, energy seems to have gone astray, but that would be contrary to the law of **conservation of energy and momentum**. It requires that the total of energy *plus* momentum remains unchanged. So what's going on? A fast-moving car causes as much damage to a brick wall as a slow-moving lorry. The vehicle's 'motion-energy', its momentum, is due partly to its speed and partly to the bulk of the vehicle, its matter.

At low speeds the rocket's engine-energy translates to motion-energy mainly as greater rocket speed. But at vast speeds increases in rocket speed are limited by light-speed. So, instead, engine-energy translates to motion-energy mainly as an increase in rocket mass: *as if it is becoming heavier.*

However, 'increased mass' *does not* mean more matter. As we saw on page 26, 'mass' isn't a measure of the amount of matter in an object, but of its 'inertia': its resistance to change speed or direction.

The relativity factor, discussed on page 38, also describes how mass increases with speed. It tends toward infinity as light-speed is approached. Since it would take an (impossible) infinite amount of energy to accelerate an object with infinite mass, this is another reason why objects with mass can't actually travel at light-speed.

OBJECTS WITH ZERO MASS TRAVEL AT 'INFINITE' LIGHT-SPEED

Objects with no mass, however, can and do travel at light-speed. This includes electromagnetic waves such as light. In fact they can *only* travel at light-speed. If Einstein could have his wish and travel alongside light from a distant star, time would appear to have stopped and the universe contracted to nothing in its direction of travel; and so the light would appear to have crossed the universe at infinite speed. That's why Einstein described light-speed as the universe's infinite speed even though it's actually a finite number.

Just as the number 'infinity' is bigger by infinity than all other numbers, so no matter at what speed her rocket travels an astronaut will always see light travelling at the universe's 'infinite' speed faster than her rocket; as the Michelson-Morley experiment demonstrated.

THE SUMMARY SO FAR

- There are no universe-wide reference points for space or time. Observers observe each other from their own reference frames.
- The universe's speed limit, light-speed, is invariant (i.e. the same for all observers, no matter their speed).
- Space is not rigid and time is not absolute.
- As an object's speed through space increases it trades off against its decreasing speed through time. As a result, as relative speed between two observers increases each believes that length has contracted, time slowed and mass increased for the other. At low relative speeds these effects are imperceptibly small.

4 A RADICALLY NEW VIEW OF THE UNIVERSE

Congratulations, You've Just Done the Hardest Part

Whilst relativity defies expectations it has been demonstrated in many ways. See the chapter, *Where's The Proof* on page 67. No other theory addresses the problems with the old physics, incorporating *both* the invariance of light-speed and the Principle of Relativity.

Things are easier to understand from here on and continue to use simple logic to reach more universe-shattering conclusions.

Introducing Spacetime

The summary on the previous page points to a relationship between space and time, one which our low-speed intuition doesn't perceive. Together they form a single four-dimensional (4D) entity called 'spacetime'[8], comprising three dimensions of space[9] and one of time. (In the graphic at right, only one of the dimensions of space is shown for simplicity.) Whilst we perceive space and time as separate, they are actually just different perspectives of spacetime.[10]

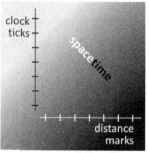

But the idea of spacetime shouldn't be so surprising. We're used to hearing that distant galaxies are a large number of light-years from us. A **light-year** is the distance light travels in a year. So if we ask, 'How does the universe look now?' we can't possibly know. We can only know how the sun looked 8.5 minutes ago, or how the edge of the visible universe looked 13.8 billion years ago. Space and time are clearly bound together in our attempts to describe the universe.

---

[8] Newton's universe was also mathematically 4D (3 space plus 1 time). However, for Newton, the dimensions of space and time were separate and not related; whereas in relativity they are related to each other.

[9] The three dimensions of space are: up/down; side to side; forwards/backwards.

[10] Consider the parable of the ***Blind Men and the Elephant***. Blind men unaware that elephants exist disagree about what they're touching. To one the trunk indicates a snake, to another the leg indicates a tree, and to the third the ear indicates a fan.

VISUALISING SPACETIME

Spacetime is not a physical thing, it's simply a mathematical model of the universe. Trying to visualise 4D spacetime is hard for we three-dimensional-thinking (3D-thinking) humans. To make a start let go of the idea that spacetime *is* the universe. Instead consider spacetime as a 4D model which *describes* the universe. Next, try to shut-down your brain's intuition which is based on its 3D mental models, otherwise it will resist trying to understand what it cannot visualise. The following analogy may help.

Challenging preconceptions: the limits of our 3D senses
Survival in a low speed world has not required humans to visualise 4D so we've not evolved that ability. Since we can't visualise 4D spacetime, it's helpful to understand the limits of our 3D intuition by analogy to a being whose intuition is limited to two dimensions.

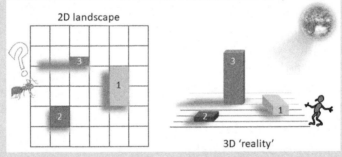

A two-dimensional-thinking (2D-thinking) ant, asked to order the blocks at left, will say: 1, 2, 3. The ant, with no conception of 'up', or physical ability to look 'up', cannot see that the objects have height, as at right in the graphic. We, 'superior' humans, able to see the third dimension of space, arrive at a different order: 3, 1, 2.

Furthermore, our 2D-thinking ant has long been bemused by the dark grey patches that appear to move around each object. Then one day a 2D-thinking Einstein says, 'What if these objects have a third dimension, let's call it *height*? And what if the day and night we experience is because of a bright light that moves in this third dimension? Maybe the height of the objects would block out the light and create dark grey *shadows* that move as the bright light moves.'

It should be clear from this analogy that unusual discoveries may lie beyond our everyday senses, especially when those senses are one dimension short of actual reality.

> Any real body must have ... Length, Breadth, Thickness, and Duration. But through a natural infirmity of the flesh ... we ... overlook this ... because it happens that our consciousness moves intermittently in one direction along the latter from the beginning to the end of our lives.
>
> *The Time Machine, HG Wells (1895)*

REPRESENTATION OF CONTRACTING LENGTH AND SLOWING TIME

On pages 16-18 we used an analogy to represent the trade-off between space and time as speed increases. Einstein's tram ride provides a real world example. Now let's see how the trade-off can be represented graphically.

Two observers look out at spacetime, each from their own frame of reference. The relativistic distortions of space and time, due to their vast speed difference, can be represented as a rotation and distortion of the other's reference frame, as shown at right.[11] Physicists call this a **'transformation'**. (See A Little Bit of Maths, page 115, for more detail on transformations.)

We see here how the observer believes that length and time in his own reference frame are unchanged, whilst appearing transformed in the other's reference frame. And, once again because of the Principle of Relativity, while this picture shows the perspective of one observer, a similar but reversed picture would show the other observer's perspective.

---

[11] This is not how spacetime *looks* but a representation of how it *works*. For simplicity, only one dimension of the three dimensions of space is depicted. This picture is a form of **spacetime diagram**. (See A Little Bit of Maths, page 109, for more on spacetime diagrams.)

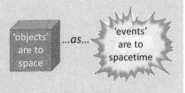
WHAT IS SPACETIME MADE OF?

Relativity invokes a single 'spacetime' model with some certainty. So presumably there's a clear idea what it's made of. Unfortunately not.

Many different concepts are used to describe spacetime: 'fabric', 'fluid', 'field' and 'grid'. But no-one has actually seen spacetime. All we know is that this *mathematical model* works at explaining how various aspects of the universe work. This lack of knowledge doesn't undermine relativity, it just highlights that it is an incomplete theory; in just the same way that we're happy to climb into aircraft designed and flown according to Newton's equations even though he didn't know how gravity worked.

There is much debate among physicists about what spacetime might be made of. A few theories are being worked on, and current or planned experiments aim to delve ever-deeper. The competing ideas all suggest that at the very smallest scale spacetime is some form of fluctuating field of energy. (See also page 83.)

Whichever the right theory it will need to answer how spacetime, whatever it is made of, assembles itself to produce the universe revealed by relativity. And then physicists will have another question to answer: what gave rise to whatever that 'stuff' is?

So, while the universe is clearly not as it seems, it seems we must continue waiting to find out what the universe actually is.

LIKELY PROPERTIES OF SPACETIME

Even though physicists know very little about what spacetime is, it is possible to consider what some of its properties might be. In so doing we discover similarities and differences between space and time.

> ➤ Space and time are not separate entities. They are simply different perspectives of a single 4D entity, spacetime. Spacetime operates differently than our 3D expectations.

> ➤ We tend to think that space enables the placing of points in space, and that time enables the sequencing of moments in time. In fact, spacetime combines these notions, enabling the arrangement of *events* in spacetime.

> ➤ We tend to imagine space as rigid and stationary and time to flow at a constant rate from the future through the present to the past. In fact, an observer's perspective of space and time is not rigid or absolute, unlike the universe which Newton had assumed. In spacetime the faster we travel in space the more slowly we travel in time, leading to length contracting and time slowing.

> ➤ If the travel of a massless object such as light is constrained to a finite speed, it suggests that spacetime may in some way be constraining its speed: an interaction of some sort between the object and spacetime. There are further consequences of such interactions, as we'll see when we discuss gravity on pages 52-55 and also in the chapter, *Theory of Everything, Quantum Matters*.

> ➤ The previous two points suggest that objects are actively engaging with spacetime. Spacetime must therefore be part of the universe and not simply the stage upon which the universe acts out its history. That implies that space and time must have come into existence with the universe, and will cease to exist when the universe ceases. We'll return to this idea briefly in the chapter, *Relativity's Philosophical Implications*.

> Time and space and gravitation have no separate existence from matter.
>
> *Albert Einstein*
> *(When forced to summarize the general theory of relativity)*

5 REVISING THE LAWS OF PHYSICS

RATTLING THE FOUNDATIONS OF PHYSICS SOME MORE

Up until Einstein came along the laws of physics governing how objects move and interact seemed set in concrete. These laws were embodied in equations which connect ideas such as force, energy and momentum to the fundamental properties of distance, time and mass. And these laws worked perfectly in a low-speed universe.

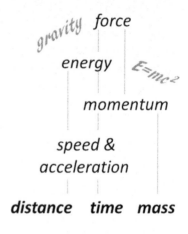

However, as we've seen, relativity demonstrated that the fundamental properties of distance, time and mass are not absolute but vary with speed; especially at vast light-like speeds. As such it appeared to be shaking the very foundations upon which the whole of physics had been built.

It was therefore necessary that the equations describing the laws of physics be modified to ensure that they worked at all speeds, including light-like speeds.

This led to more expectation-shattering insights and heralded the dawn of the nuclear age. On the next pages we will look firstly at the relationship between 'mass' and 'energy' as described by $E=mc^2$. Then we will look at 'force', specifically what relativity has to say about how the force of gravity actually works.

> A man should look for what is, and
> not for what he thinks should be.
> *Albert Einstein*

Mass-Energy Equivalence and the Famous Equation E=mc^2

Arriving at E=mc^2

We saw earlier that as a rocket already travelling at vast speed fires its engines to try to go even faster, its engine-energy converts to rocket motion-energy through an increase in its mass instead of its speed. As we saw on page 26, 'mass' is simply a property of matter: a measure of its inertia and not simply a measure of the amount of matter it contains. As such, a change of mass can result from a change in many factors, see page 51.

The fact that mass can convert to energy was not a totally new insight. Other examples were known by the time Einstein came along. For example, as radioactive materials such as uranium decay they not only radiate energy but they also lose mass (in this case in the form of matter). The question that was coming into focus was, "What exactly is the relationship between energy and mass?"

An object in motion has 'motion-energy', or **momentum**, as a result of its motion. That momentum is a function of its mass and speed. It may seem a surprise to discover that electromagnetic waves such as light have momentum *despite* having no mass at all. In other words, light can behave just like a colliding snooker ball. It was Maxwell's theory of electromagnetic waves which first suggested this idea, and which was demonstrated experimentally in 1923.

Einstein saw in this an opportunity to derive the mathematical relationship between energy and mass by considering how a massless light beam would transfer its momentum to an object with mass.

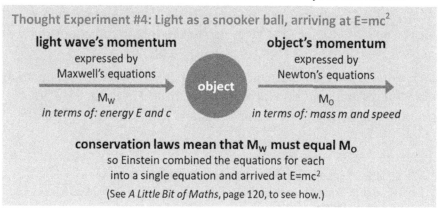

Thought Experiment #4: Light as a snooker ball, arriving at E=mc^2

light wave's momentum
expressed by
Maxwell's equations

M_W
in terms of: energy E and c

object's momentum
expressed by
Newton's equations

M_O
in terms of: mass m and speed

conservation laws mean that M_W must equal M_O
so Einstein combined the equations for each
into a single equation and arrived at E=mc^2

(See *A Little Bit of Maths*, page 120, to see how.)

What Does $E=mc^2$ Mean?

It is not intuitive that we should arrive at such a simple equation as $E=mc^2$ but that's how the maths falls out.[12] (See *A Little Bit of Maths*, page 120.) It is this simplicity which makes $E=mc^2$ so important. The equation makes no reference to how the object is moving. In fact, it makes no reference to anything other than 'E', 'm' and the universe's constant, 'c', the speed of light.

Einstein inferred from this that $E=mc^2$ was not confined to light beam's striking objects, but that it could be generalised to all forms of energy.[13] He expressed this idea of so-called '**mass-energy equivalence**' in his second 1905 paper on relativity[14] as follows.

> '*If a body gives off … energy … in the form of radiation, its mass diminishes [and] becomes energy of radiation … so that we are led to the more general conclusion that … the mass of a body is a measure of its energy-content …*'

We know that energy can be converted from one form to another, for example electricity to light. But, Einstein's idea was startling: if 'E' and 'm' are simply related by the universe's constant, 'c', then all sources of mass even solid matter are 'just' energy. $E=mc^2$ tells us that densely-packed energy takes on the properties of matter. Physicists use the term '**energy density**' whether referring to less dense 'energy' or more dense 'matter'.

This suggests that matter can be completely converted into energy, and not just partially through chemical reactions such as burning. Since 'c' is a huge number (300,000,000 m/s), so c^2 is enormous, nearly 10^{16} (10 followed by 16 zeros). This vast amount of energy locked up in matter is what gives rise to the threat and opportunity of nuclear reactions.

Einstein's generalisation also means that if we provide energy to an object it will appear as an increase in its mass. For instance, a hot drink will be heavier than a cold drink and its inertia will be greater.

---

[12] The equation $E=mc^2$ was initially written differently by Einstein.

[13] Einstein didn't prove the generalization, he asserted it as an implication of $E=mc^2$.

[14] A paper entitled, '*Does the Inertia of a Body Depend Upon Its Energy Content?*'

Energy comes in different forms, each part of a different process. There are those which we can observe at human scale, going right down to the inner workings of atoms. Any increase to an object's energy increases its energy density and so results in increased mass, or inertia – *as if* it had more weight.

Kinetic Energy

- The energy an object has due to its relative motion
- It is increased or decreased by changing speed

Thermal energy

- The energy atoms or molecules have due to their vibration
- It is increased or decreased by heating or cooling

Chemical energy

- The energy in the bonds between atoms in molecules
- It is increased or decreased through chemical reactions

Electromagnetic energy

- The energy of electrons orbiting atoms; or in electromagnetic waves
- It is increased or decreased by absorbing or radiating energy

Nuclear energy* (approximately 1% of an atom's mass)

- The energy which binds protons and neutrons in an atomic nucleus
- It is increased or decreased by nuclear reactions or radioactive decay

Fundamental particle energy (approximately 99% of an atom's mass)

- The densely-packed energy in sub-atomic particles
- It is increased or decreased by the matter-antimatter pair production or annihilation of particles

(NB **Potential energy** isn't a separate energy form. Instead it expresses the potential for energy to be released. It is the difference between an object's or a particle's current state relative to a lower energy state.)

* Nuclear energy, nuclear force and gravity
Nuclear forces are very strong, up to 10^{38} times stronger than gravity. Their nuclear energy binds together the components in the atomic nucleus many of which, having positive charge, are trying to repel each other. Despite the strength of nuclear forces, we never experience them directly since they have an exceptionally small range.

Gravity, on the other hand, is a much weaker force but it does have exceptionally long range. When applied to massive objects like planets it becomes a significant force on cosmological scales.

(For more on forces see the Standard Model on page 81.)

AN EXPLANATION FOR GRAVITY

Gravity and acceleration are indistinguishable

Continuing his review of the laws of physics, Einstein turned his attention to gravity. Newton had provided very accurate equations by which to calculate gravitational attraction between two bodies moving at everyday speeds. But he had not explained how gravity actually worked: from where that 'force of attraction' came.

Einstein realised with the help of a further thought experiment that gravity might not actually be a force of attraction at all.

Thought Experiment #5: 'The happiest moment of my life'
Einstein described as *'the happiest moment of my life'* this inspiration. *'Suddenly a thought struck me: If a man falls freely, he would not feel his weight. I was taken aback. This simple thought experiment made a deep impression on me. This led me to the theory of gravity.'*

Einstein had realised that a man falling from a building and a man in a rocket floating freely in space each experience weightlessness in exactly the same way. Indeed, if both are blindfolded in a spacesuit and suddenly awakened from sleep, they'd have no way of knowing which situation they were in – just as Galileo's sailor in his cabin led Galileo to the Principle of Relativity.

And when the falling man reaches the ground (hopefully on his feet), or when the rocket starts to accelerate, the sensation, the pressure felt between feet and floor, will also be identical for both.

Einstein called this the '**Equivalence Principle**', meaning that it not possible to distinguish between acceleration and gravity.

Einstein inferred from this Equivalence Principle that gravity was 'just' an effect, the result of objects accelerating through spacetime. But if that is the case, what is it that creates that acceleration?

Matter causes spacetime to curve ...

To explore this question further, Einstein wondered how acceleration might affect the trajectory of light beams.

Thought Experiment #6: If acceleration bends light so too must gravity
A light beam shines through the windows of a rocket. Depending on if and how the rocket is moving relative to the light source, the beam appears, *to the rocket's astronaut*, to take different trajectories.

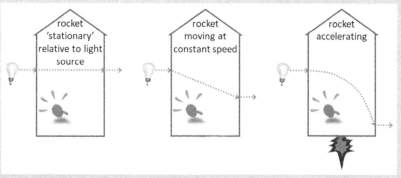

If the rocket is accelerating (i.e. it's speed is increasing) the light's trajectory will appear curved to the astronaut. But, according to the Equivalence Principle (page 52), if the path of a light beam appears curved by acceleration, it must likewise, indistinguishably, follow a curved trajectory in the presence of a gravitational force. This is a surprise: we expect light beams to travel the shortest distance between two points. So what's going on?

It is of course true that the shortest distance between two points is a straight line *on a flat surface*. However, on the surface of a three-dimensional object the shortest distance is a curve. For example, the shortest distance from Paris to Los Angeles is not straight across the Atlantic but over Greenland.

Let's apply this insight from 3D geometry. If the light beam's path through spacetime is being curved by the presence of a gravitational force, it suggests that spacetime is being curved by the object creating that gravitational force. Or, put another way, if gravity curves the path of light then gravity must be the result of curves in spacetime.

(See *A Little Bit of Maths*, page 123-5, for more on this.)

We've described spacetime as a four-dimensional entity. That's a little vague. It's more helpful to think of spacetime as similar to a magnetic field. Indeed, physicists talk about the **'spacetime field'**.

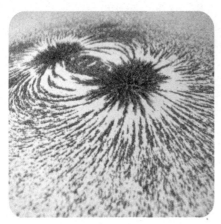

Science experiments use iron filings to 'observe' the magnetic field around a magnet. The image at left shows how the magnetic field lines are curved by two metallic objects. In other words, the metal objects have in some way interacted with the magnetic field. They in turn become magnetised and as a result pull towards each other.

Similarly, the densely-packed energy which makes up objects interacts in some way with the spacetime field. This results in a concentration of the spacetime field in the object's vicinity, creating a 'gravity well' (or pressure gradient) which other objects experience as gravity.

... and curved spacetime tells matter how to move
Just as metal objects are influenced to move by magnetic fields, so all objects respond to the curves in the spacetime field. That's why the spacetime field is sometimes known as the **'gravitational field'**.

We saw earlier that it is impossible for us as 3D-thinking humans to fully visualise spacetime. But there are some visualisations which can help us understand the behaviour of a spacetime field.

In the visualisation at right we're looking at a slice of spacetime. We can imagine this slice painted onto a rubber sheet. The depression caused by the earth can be represented by grid lines mapping out the resulting 'pressure gradients' in the spacetime field.

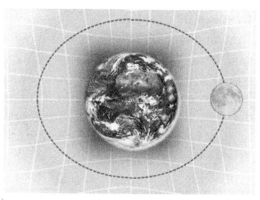

Matter tells spacetime how to curve and curved spacetime tells matter how to move.

John Wheeler

The effect is that any nearby objects are drawn towards each other, just as two people on a mattress might roll towards each other.

In other words, there is no separate force of gravity; it is 'just' the result of an object responding to the pressure gradients in spacetime.

Objects can remain in orbit rather than roll down the curves of the gravity well. This would be the case if they're travelling at speed around each other, balancing a tendency to fall into the gravity well with an opposing tendency to fly off into space. Similar mechanisms keep the moon, planets and stars in their orbits.

The problem with conceptual visualisations of spacetime
Whilst visualisations of spacetime can helpfully explain part of a concept they can also create false impressions. For example, we might think that it is natural for an object to depress a rubber sheet because of gravity pulling from below. But, pausing to reflect, we realise that:
(1) Since the analogy *explains* gravity, there is no gravity acting from below; especially as it is set in deep space and not on earth.
(2) It is not known exactly *how* matter interacts with spacetime to create the impression of mass and cause spacetime to curve. (The '**Higgs field**' plays a part, imparting the property of mass to objects, see page 82.)

Furthermore, the grid lines suggest that spacetime is fixed in place in some way. But as we'll soon see, spacetime, i.e. the universe, is actually expanding all the time.

BENDING OF STARLIGHT

The bending of light we saw in Thought Experiment #6 can also be explained through mass-energy equivalence which we discussed on page 50. There we explained that physicists often use the term 'energy density' whether referring to less dense 'energy' or more dense 'matter'. Since light is a form of energy, for example giving life to plants and animals or causing burns, it has an energy density despite having no mass.

As such, we should expect starlight to be deflected as it passes through gravitational fields. This was the basis of one of the most important early proofs of relativity. (See *Where's the Proof?* page 67.)

BLACK HOLES AND SINGULARITIES

If the object causing the curvature of spacetime is sufficiently massive, spacetime curves away into a so-called **'black hole'**. The energy density of a black hole is so great that it creates an infinite curvature of spacetime at its centre, like a bottomless well in spacetime's fabric. This is called a **'singularity'**, and any matter or energy (such as light), once drawn in, cannot escape.

GRAVITATIONAL REDSHIFT, GRAVITATIONAL TIME DILATION

Imagine light trying to leave a massive object, such as a star. Two things will happen to the light beam.

As we've just learned, light is influenced by gravity. The object's gravitational influence will try to pull the light beam back, causing the light's wave to be stretched. To an observer, this longer wavelength light will shift the light's colour toward the red end of the spectrum. This is called **'gravitational redshift'**. (A more technical explanation is that the light must expend energy to escape the gravitational field. As the light's energy decreases, its frequency decreases – which by definition means a shorter, redder, wavelength.)

Furthermore, since the geometry of spacetime dictates that the light beam travels a curved path out of the gravity well in which the object sits, that path will be a longer path. But relativity says that the observed speed of light must be the same for all observers.

Therefore time, for an observer near the massive object, must be slower than for one further away if he is to arrive at 'c' when calculating the light beam's speed. In other words, time ticks at different rates whenever there is a difference in gravity between two points. This effect is called '**gravitational time dilation**' to differentiate it from time dilation caused by relative speed difference. In a black hole from which light cannot even escape gravitational time dilation is so great that time actually slows to a complete standstill.

An example much closer to earth is that of satellite navigation systems and the orbiting GPS satellites on which they rely. The clocks in the satellites need corrections to ensure navigational accuracy on earth. It's not especially because the orbiting satellites are travelling at great speed relative to the earth. It is mainly because they are in a significantly weaker gravitational field than the clocks built into our sat-nav devices here on earth with which they must stay aligned.

Nothing Except Empty Curved Space

Relativity reveals two previously unknown deep relationships: the one between mass and energy, and the other between space and time. We have also seen that these two relationships are themselves connected: '*Matter tells spacetime how to curve and curved spacetime tells matter how to move*'. John Wheeler, who gave us this last quote, put the implications of relativity in a nutshell.

> There is nothing in the world except empty curved space. Matter, charge, electromagnetism, and other fields [such as gravity] are only manifestations of the curvature of space.
> *John Wheeler, Physicist*

Gravitational Waves

Amongst Einstein's predictions arising from relativity was the existence of '**gravitational waves**', often described as 'ripples in spacetime'. To understand these imagine a frictionless sphere spinning 'on the spot' in water. Because the sphere is frictionless, the water is undisturbed. But, if the rotation is uneven or the sphere suddenly removed altogether, the water is disturbed and waves will ripple outwards.

Likewise gravitational waves result from disturbances of spacetime. For example, 'binary stars' (a pair of massive stars orbiting each other every few minutes) whip up spacetime's fabric creating strong gravitational waves.

Since nothing can travel faster than 'c', the gravitational waves are also limited by this universal speed limit. If our sun suddenly vanished this would constitute quite a disturbance of spacetime. It would, of course, take 8.5 minutes for us to notice the light disappear from the sky: that's the time it takes for light from the sun to reach the earth. It would also take 8.5 minutes for the earth to register the absence of the sun's gravity and fly out of its orbit into deep space. That's how long it would take the gravitational wave disturbance to reach us.

Since everything in the universe is woven into the fabric of spacetime any movement can create gravitational waves. These waves cause objects they pass through to bend and stretch in response, rippling them. If the effect of these ripples could be seen with the naked eye, objects would look like those reflected in the rippling surface of a lake.

For the most part however, these ripples in spacetime are so weak in our calm backwater of the universe called the solar system that they're barely detectable. However, experimental evidence for gravitational waves emanating from a very distant but hugely powerful cosmological event was announced on February 11th 2016, thereby becoming the last of Einstein's predictions to be confirmed.

6 INSIGHTS INTO THE UNIVERSE'S DESIGN

COLLAPSING UNIVERSE ... OR ... EXPANDING UNIVERSE?

There is nothing much within Einstein's model of the universe that prevents gravity from causing the whole universe to collapse in upon itself. The mutual gravitational attraction of the universe's matter and energy should cause it to collapse like a deflating balloon.[15]

There is only one workable reason it wouldn't do so, and that's if it was forever expanding. Even if at some point the universe had achieved some sort of stability, it would take only a small disturbance (for instance a collision between galaxies) to introduce instability leading to collapse or expansion; much like a pencil, finely-balanced on its end, topples if nudged even slightly.

But in the early 1900s astronomers looking into the night sky with telescopes then at their disposal, observed a universe that looked static and not one that was collapsing or expanding. So Einstein adjusted his equations to counter-balance this theoretical tendency to total collapse, enabling his equations to describe a static universe.

However, by 1929 conclusive evidence for an expanding universe had been found. As the universe expands, its galaxies are being carried ever outward by the tide of expanding space and time. So Einstein did away with his adjustment, and the search began for the mechanisms driving the universe's expansion.[16]

THE SHAPE OF THE UNIVERSE

Even though spacetime is curved locally by planets, galaxies, black holes, etc., it is thought that overall the universe is on average not curved but 'flat'. And also, paradoxically, that it is finite in size and yet unbounded by any edge! (See page 130 for more detail on this.)

---

[15] Indeed, this would also be true of Newton's universe, except that he assumed the universe was infinite, thereby avoiding its theoretical collapse.

[16] This adjustment called the **'cosmological constant'**, has reappeared in recent decades as cosmologists try to explain how our universe can be ever-expanding by incorporating new ideas about **'dark energy'**. (See *A Little Bit of Maths, page 129*.)

A Big Bang

If the universe is expanding, it suggests there was a beginning to the expansion. And since by 'universe' we mean spacetime then, at that beginning, space is shrunk to nothing, and time did not exist. (If this sounds similar to the black hole singularities we met earlier, you're right, it is.)

From here, some theories suggest, the universe came into existence through random fluctuations in energy fields followed by an explosive inflation (see page 83). (Alternative theories propose that the Big Bang is actually a rebound, that is, a 'big bounce', back from a previous collapsing universe.)

'c' is Very Special

Why can't objects travel at the speed of, or faster than, light?
Neither rockets nor other objects with mass can travel at light-speed. As we've seen, it would require infinite energy to accelerate them to that speed. It's not because there is anything special about light. It's just that being mass-*less* it can travel at the fastest speed the universe allows. It is this universal constant, 'c', which is special. And of course, all other (equally massless) electromagnetic waves can also travel at this same light-speed.

However, whilst Einstein's equations indicate it's impossible to travel faster than light in a vacuum, we should bear in mind that scientists used to argue that man-made objects couldn't travel faster than sound. As ever, it's best to never say 'never'. Some scientists are today considering how faster-than-light travel might one day become a reality.

Designing the universe
What we've discovered is that the universe is defined by fixed parameters, but these are not the properties of distance and time. It is instead defined by constants such as the speed of light – one of nature's design parameters. Not only is it the universe's speed limit but it also defines the relationship between space and time, and between mass and energy.

7 A FEW LOOSE ENDS

WHAT DOES 'RELATIVITY' ACTUALLY MEAN?

The word 'relativity' refers to the Principle of Relativity. As we have seen this tells us that we can only describe the position or motion of an object *relative* to other objects. The *laws of physics* relating to things such as distance, time and mass apply identically to all frames of reference, just as the Principle of Relativity demands. But the *values* observed for these things depend on relative speed.

> **What's in a name?**
>
> Einstein regretted the name 'relativity' because of inappropriate popular use. It led some to understand that everything was relative and varied with speed, even the laws of physics. But, as we've seen, the laws of physics are *absolute* and always operate in exactly the same way. It is only the properties upon which the laws operate which vary with speed, properties such as distance, time and mass.
>
> But Einstein's regret at the name 'relativity' extended beyond misunderstood science. Relativity was taken up by proponents of '**Relativism**' to support their belief that there is no absolute truth or validity for anything, for instance as applied to morals. Whilst arguments can be made for Relativism, relativity is at best a very poor example since, in actual fact, it presumes that the laws of physics are actually, and always have been, absolute.

'SPECIAL' AND 'GENERAL' RELATIVITY

Einstein's two 1905 papers on '**Special Relativity**' applied to objects moving relative to each other with unchanging speed and direction. It was 'Special', because it applied only to this special case.

Understanding the relativistic effect on objects which were accelerating, that is changing speed or direction, took Einstein a further ten years. He first presented this work in 1915 and published his final paper on '**General Relativity**' in 1916.[17] This was 'General' because it applied to all objects however they are moving.

---

[17] Einstein's work on General Relativity included many papers from 1906 to 1916.

If All This is True, Why Don't We See Any of It?

Let's go back to the earlier analogy on page 10 where we grew up on a gently rolling plain, as far as the eye could see, with no knowledge of huge mountains or vast oceans. In a similar way, our planet earth is in one of the many calm backwaters in an elsewhere tempestuous universe. In these calm backwaters there are no truly massive objects, nor objects travelling at vast light-like speeds relative to each other.

We should be grateful. Being caught up in such tightly curved regions of spacetime, for instance around black holes, would be far worse than being caught in the middle of a raging sea. Life as we know it could not exist in such regions. Enormously powerful gravitational waves would create equally enormous differences in the gravity felt by different parts of our bodies. These differences would tear us apart, not just limb from limb but atom from atom.

It is only with the development of far more sensitive instruments and telescopes that our technology-assisted human senses are increasingly able to detect the effects of relativity.

How Wrong Was Newton?

It might seem that Newton, or at least his theory, comes out of this story somewhat tarnished. That would be an incorrect perception. He is rightly hailed as one of the scientific greats because of his achievements. His theory, understandably, could not account for phenomena which in his day had not even been observed.

> No one must think that Newton's great creation can be overthrown in any real sense by this [theory of relativity] or by any other theory. His clear and wide ideas will forever retain their significance as the foundation on which our modern conceptions of physics have been built.
>
> *Albert Einstein*

And, even more than that, we can say that Newton was right: for a *matter-based* view of the universe. Newton's Laws are still used today, serving perfectly well for most everyday low-speed purposes.

Einstein's relativity describes a richer, more inclusive, *energy-based* view of the universe. One underpinned by the invariance of the speed of light and the Principle of Relativity and which encompasses and extends beyond Newton's matter-based laws.

What Michelson-Morley's 'failure' tells us about scientific research
It seems ironic that the Theory of Relativity followed Michelson-Morley's 'failed' attempt to find evidence for the ether, the presumed medium for transmitting light. The irony is that it contributed, after all, to such important advances in our understanding of the universe.

Michelson and Morley were highly respected scientists. The fact that their experiment did not achieve its aim was possibly a disappointment to them. They were not to know in 1887 that this 'failure' would help unlock a new understanding of the universe, contributing to the development of important theories as well as practical systems upon which we rely every day.

It's a demonstration of the importance of basic scientific research, and that we can learn from our 'failures' as well as from our successes.

YOU MADE IT!

Congratulations, you have come to the end of the explanations of relativity.

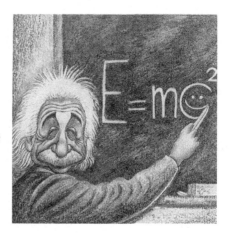

You should feel good about that even if you feel your grasp is in some way incomplete. You would not be alone in taking time to absorb the many broad and challenging concepts within the theory of relativity.

Let's remind ourselves quite how broad and challenging it is.

A Summary of Relativity's Startling Conclusions

- The underlying fabric of the universe can be represented mathematically as four-dimensional spacetime. Space and time are not rigid or absolute properties. There are no universe-wide reference points from which everything is measured. (We may not know what spacetime looks like, but we can model how it works, just as Newton modelled how gravity worked without knowing what it was.)

- The separateness of space and time which we experience arises from our limited perceptions (narrow glimpses) of spacetime. Space and time are related to each other.

- Our sense that the universe is defined by the properties of distance and time is wrong. It is instead defined by other parameters, nature's constants, such as 'c'. 'c' is the universe's speed limit, the speed at which massless electromagnetic waves such as light travel; and the maximum speed at which information can be conveyed.

- As moving objects approach this speed, relative to observers, it appears to the observers that the object's length has shortened, time for the object has slowed, and its mass has increased.

- Energy and matter are equivalent, related one to the other by 'c'. Matter is just very high density energy.

- The presence of matter or energy causes spacetime to curve, creating pressure gradients. These pressure gradients result in the force which we experience as gravity.

- The universe, that is spacetime, is expanding carrying all energy and matter with it as it does.

In short, the relationship between 'c', space, time, energy and mass is inextricable. It is cemented into the fabric of the universe. It is how the universe is built.

NEXT TIME YOU GAZE AT A CLEAR NIGHT SKY

Next time you gaze up at a clear night sky, stop.

Look deep into the Infinite and picture Einstein's Universe; a gently bubbling ocean of Energy, encased in pure Darkness.

Here and there more densely-packed in vast clouds of gas or even denser galaxies; at times so compressed as to fuel suns and form planets which bring forth Light and Life. And again denser still, such that not even fleeting Light can escape the Blackness.

Peer deeper still. Now imagine the ocean of Energy woven into the all-enveloping fine silk of Space and Time; itself curved to greater or lesser degree according to the density and influence of Energy: Gravity's hills and valleys, thus created, gently guiding the graceful Motion of the Heavens.

And know, in that moment, that 'c' the universe's Infinite Speed runs its mathematical rule on all this majesty.

WHERE'S THE PROOF?

This chapter describes the significant physical evidence for relativity, and also discusses what 'scientific truth' really means.

RELATIVITY PROVEN ...

The test of a good theory is in three parts. Does it: explain what we know, does it make valid predictions, and does it help us build things that work? Here's a summary of relativity's proofs.

- Relativity **explains Michelson-Morley's finding** that the speed of light is independent of an observer's relative motion.

- Relativity explains why the **observed orbit of Mercury** is subtly different to that calculated by Newton's Laws. The finite speed of gravitational influence means that at any instant, Mercury is attracted to where the sun was a moment before. This causes the orientation of its elliptical orbit to constantly shift. Such so-called '**precession**' affects all planets, but Mercury's small size and proximity to the sun means its precession is pronounced.

- The **bending of starlight** by massive objects was a predictable result of Newton's Laws. However, only relativity precisely predicts the extent of the bending, as experimentally verified in 1919. (The degree of precision of the 1919 result has been questioned, but the proof has been repeated and confirmed many times.)

- **Gravitational lensing**, commonly used by astrophysicists, uses the bending of starlight by gravity to magnify distant objects.

- Relativity's equations support the observation first made in 1929 that the **universe is expanding**.

- The opportunity and threat arising from mass-energy equivalence and observed in **nuclear reactions**.

- The **gravitational redshift** of light from a massive burned out star, known as a white dwarf, was verified in 1959.

- The first **black hole** was discovered in 1971.

- The **proof of time dilation** (of time slowing). In 1971, two highly accurate atomic clocks (able to measure time to a ten-billionth of a second) were flown in opposite directions around the world. On their return they were compared with a stay-at-home clock. They had drifted apart by an amount matching relativity's predictions. This time dilation was in part due to relative speed and partly due to gravity (the aircraft were obviously at higher altitude, and so their clocks were less affected than the stay-at-home clock.)
- Relativity explains the **unexpected behaviour of muons**. These particles, created by cosmic rays colliding with the earth's upper atmosphere, have a very short half-life and should decay before reaching the earth's surface. But muons travel at near light-speed, so the distance they need to cover to the earth's surface is significantly contracted. As a result they reach ground detectors more quickly, many more than otherwise expected doing so before decaying.
- The 2016 experimental proof that **gravitational waves exist**. (Einstein thought these might be too weak to measure on earth.)
- **Body scanners and particle accelerators** must control the path of particles moving at 99% of light-speed, or more. Designers of these systems must account for relativistic increase in their mass.
- The design of **satellite navigation systems** must also allow for **gravitational time dilation** otherwise significant discrepancies would appear between actual position and that given by sat-navs.

... BUT NOT FOR SOME PEOPLE ...

The Theory of Relativity was controversial for many years. The science was complex and its implications contrary to many people's beliefs. Much pseudo-science tried to refute it; a minority motivated by religious conviction and some, primarily in inter-war Germany, by anti-Semitism toward Einstein (see the chapter, *Relativity, Culture and Society*, page 98). Einstein himself despaired.

> *'This world is a strange madhouse. Every coachman and every waiter is debating whether relativity theory is correct. Belief in this matter depends on political affiliation.'*

In the early 1900s, relativity was too controversial to award Einstein a Nobel Prize. So instead the Nobel Committee, impatient to recognize Einstein, awarded him the 1921 Nobel Prize for Physics for his other ground-breaking work of 1905 on photo-electricity.

Relativity continued to be hotly debated in scientific circles well into the 1920s. Some serious and well-intentioned scientists questioned relativity theory's assault, as they saw it, on the very foundations of physics. Today, relativity is far less controversial but remains little understood outside parts of the scientific community. Nonetheless, engineers and scientists go about their business relying on relativity's equations, and we in turn rely on the systems they build. Despite this, there is still a tiny but active minority of doubters, who lie outside the mainstream of scientific opinion.

... And Not Forever

Just as Newton's theory eventually failed to explain everything, so too will the time come for relativity to be amended. And, just as relativity encompassed Newton's Laws, it is likely that any new theory will encompass (or amend) rather than replace relativity.

Ironically, it may turn out that Einstein himself set in motion the finding of evidence which relativity fails to explain. In predicting the existence of gravitational waves he set physicists the challenge of building instruments to detect them. Now these instruments, **'gravitational wave telescopes'**, exist, enabling astrophysicists to probe even further into the universe and deeper into its history. Just possibly it will be here in these echoes of the universe's earliest moments, or in its most massive and turbulent objects, here where our known laws of physics are most tested, that relativity will be found wanting.

69

Is Relativity True?

Science is not a search for 'truth' but for models which explain what we observe. The more we learn, the more the models evolve.

Most people who believe in Newton's theory don't understand the mathematical calculus he invented to explain how planets move. But they believe in some notion of gravity and orbits because (they are told) it works. So why should we doubt relativity, just because it astonishes or because we don't understand the maths? It would be like saying to a heart surgeon that before he or she can replace my diseased heart, I need to understand the biochemistry, physiology and operating procedures upon which he or she relies.

So is relativity a question of faith? After all, you might say you need faith in your heart surgeon. But this is a different faith from religious faith. The one founded on personal conviction, the other on reproducible evidence and independent validation.

The Greeks started the serious search for explanations of how the world works mostly conducted as philosophical inquiry, that is thinking about things. 3,000 years later, evidence and the rigour of the scientific method came to the fore.

	Religious belief	Philosophical inquiry	Scientific method
is a search for ...	Truth	Meaning	Working models
described by ...	Stories & doctrines	Ideas & arguments	Theories & proofs
& founded on ...	Faith & personal revelation	Reasoning & independent agreement	Reproducible evidence & independent validation

It's interesting to note that Einstein's many insights were the result of '**thought experiments**', a sort of philosophical inquiry, and for which scientific methods then found the supporting evidence. It is an approach which scientists probing the cosmic and quantum worlds continue to use.

Chapter
THEORY OF EVERYTHING, QUANTUM MATTERS

To fully appreciate relativity, it is helpful to see it in the wider context of modern physics. This chapter discusses how relativity forms a part of physicists' on-going search to explain how the universe works. Their ultimate aim is to find a single, unifying, 'Theory of Everything'.

We've seen how Einstein's relativity encompassed the work of Newton and Maxwell, and redefined our understanding of space and time for large scale objects. A Theory of Everything must combine this knowledge of the large scale world with an understanding of how things work at the microscopic scale of atoms and even tinier particles.

Newton, 1687
Gravity:
Motion in space
Motion on earth

Maxwell, 1861
Electricity
Magnetism
Light

Einstein, 1905
& 1916
Time
Space
Mass

Relativity had touched on this by demonstrating that matter comprises very dense energy. That's what $E=mc^2$ tells us. But it didn't explain how that energy is organised such that the material world appears the way it does. The physics of the small scale is known as '**quantum mechanics**'. The word '**quantum**' has the same root as the word 'quantity' and means 'a fixed amount of energy'.

The challenge, however, in appreciating quantum mechanics is not in following the logic, but in its astounding implications.

> Anyone who is not shocked by quantum theory has not understood it.
> *Neils Bohr, Physicist*

Quantum mechanics adds significantly to the search for a Theory of Everything. However, it also adds to relativity's challenge to our intuition: unreality is about to become even more surreal. So keep in mind, as we start this exploration, that without this quantum weirdness the electronics that we rely on in our everyday lives would not work, to name just one thing among many.

BEFORE WE START, A FEW SIMPLE CONCEPTS

Interference, or how to add up waves
We know how to add up numbers. Adding waves is almost as easy.

At right, the top two waves are in step. When they meet they **'interfere constructively'** to form a 'tall-wave' whose height equals the other two added together, as the third wave shows.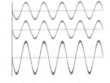

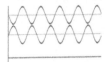 At left, the top two waves are out of step. When they meet they **'interfere destructively** and the result is a 'flat-wave'.

Two pebbles dropped into calm water produce ripples. These two radiating waves interfere with each other constructively and destructively, as at right. The result is an alternating pattern of tall-waves and flat-waves. This shows up as stripes at the right-hand edge. It is a signature pattern produced only by interfering waves.

Mathematical descriptions and interpretations, again
We have already discussed this on page 9 where we saw that while Newton did not know what gravity was, it did not impede the use of his wonderful equations. Mathematical descriptions and interpretations are highly relevant to the story of quantum mechanics, so another example will be useful.

Consider a series of numbers from 1 to 10. We eventually discover it is a random series. With this mathematical description we can now direct our search for how they were produced. Since it is random we rule out, for instance, simple arithmetic operations. Focussing on the fact that it is a random set we put forward possible explanations. It might be the number of heads from many sets of 10 coin tosses, or it might be numbers drawn from a hat. These are just possible 'interpretations' of the mathematical description. Research can now be directed to prove or disprove each possible explanation.

QUANTUM MECHANICS OVER A CUP OF COFFEE
Is it a wave or is it a particle?
In 1861, Maxwell had shown that light was an electromagnetic wave, as discussed on page 28. On the other hand, in 1905, Einstein had provided Nobel Prize-winning proof of light as particles. He theorised these as packets of light energy. (The term **'photon'** for these packets came into use from 1926, derived from the Greek word for light.)

> It seems to me that the observations associated with ... the emission or transformation of light [such as photo-electricity] are more readily understood if one assumes that the energy of light ... is not continuously distributed ... but consists of a finite number of energy quanta* which are localized at points in space, which move without dividing, and which can only be produced and absorbed as complete units.
>
> Albert Einstein
>
> * 'Quanta' is the plural of quantum.

By 1923 experiments and mathematical theory had confirmed that light sometimes behaves like a wave and sometimes as a particle. Unfortunately, Maxwell's wave theory of light could not explain light's particle-like behaviour nor Einstein's ideas explain its wave-like behaviour. The solution to this **'wave-particle duality'** led directly to a new branch of physics, quantum mechanics.

It soon became clear that other things exhibited this weird double life, for example, electrons. But electrons are one of the building blocks of the atoms which make up solid matter. How could the tiny indivisible particles which make up solid atomic matter sometimes behave like a far from solid wave?

It also became clear that in some experiments light or sub-atomic particles exhibit wave-like behaviour while in others they exhibit particle-like behaviour. It was as if they somehow chose how to behave according to the type of experiment in which they were being observed.

The 'double-slit experiment' is just one of many which demonstrate this schizophrenic behaviour. See box on next page.

This experiment demonstrates two things.

1 Wave-particle duality
When a narrow beam of electrons is projected through a single slit at a screen, unsurprisingly, we see a single white stripe on the screen. So when the electron beam is fired at two slits we would expect to see two stripes on the screen, one corresponding to each slit.

Instead what appears on the screen is the many striped signature pattern of interfering waves. (The stripes are on the far right of this picture, the two slits are to the very left.) We can't see the wave interference in the centre of the picture with our eyes but we must infer its presence from the signature pattern, and conclude the electrons are behaving as waves.

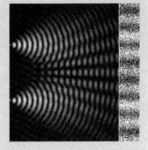

2 A particle can be in two places at the same time
Let's now slow the experiment right down so that, instead of an electron beam, just one electron at a time is fired at both slits. We do this by turning down the power in the source until, like a dripping tap, it has just enough energy to release one drop, one electron, every second. In this situation we might expect a different result.

What we see, however, is shown in the time progression of screen images at right. Individual electrons hit the screen each leaving a particle-like point, but significantly, *in many different locations.* As time progresses these points build up exactly the same signature interference pattern of stripes.

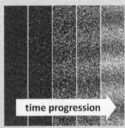

time progression

In fact, these stripes are exactly the same as the stripes in the first picture: when you look closely those are also made up of individual points. This must mean that each electron is somehow passing through both slits at the same time and then somehow interfering with itself.

In other words, each electron starts its journey from the source as a particle, behaves as a wave as it moves through the slits, and reappears as a single particle at the screen. Even trying to observe through which slit the electrons passed changes the result: if we suddenly close one slit the pattern immediately reverts to the single stripe of a particle beam as opposed to the many-striped signature image of a wave!

A mathematical description of wave-particle duality
So how can particles behave like waves, and vice versa? And how can one particle be in two places, at the two slits, at the same time? Confused, physicists looked for a mathematical description of the observations. They'd worry about the physical interpretation later.

They developed an equation, called a '**wavefunction**', which describes the likelihood, or *probability,* of a particle appearing at a particular point in space. When a particle isn't interacting with anything else the wavefunction gives it a chance of appearing in many different places in space, albeit with differing degrees of probability.

With this abstract description, the particle is wave-like, that is possibly in many places, or '**states**', at the same time. This is known as a '**superposition**' of states.

However, the moment the electron interacts[18] with something such as another particle (or indeed particles in the form of an observer) its wavefunction is said to '**collapse**' to a particle-like state in a specific place in space. Let's apply this idea to the double-slit experiment. When the particle sets off we know its location, likewise when it arrives at the screen. Unobserved between these two moments it is possibly in many places at the same time. It is this wave-like state of many possibilities which passes through both slits leaving the signature interference.

This effect is not restricted to the sub-atomic scale: in recent years the double-slit experiment has been repeated with larger objects, even some just visible to the naked eye. This is weird. How can something be wave-like one instant and particle-like the next, or vice versa, choosing between these two options depending on if and how it interacts with some other thing?

Clearly interaction between particles is an important part of this theory. But why? And what exactly constitutes an interaction? (We'll return to this idea of interactions as 'events' in the chapter, *On 'Time'.)*

---

[18] Physicists often use the word '**measure**' in this context meaning: to interact.

Uncertain times

The wavefunction tells us that a particle may be in one of many places at the same time. Even after wavefunction collapse our electron is not a precise point but a section of the probability wave confined to a narrow space: particle-*like*. This leads to the '**Uncertainty Principle**': that at any one moment we cannot measure with precision all of a particle's properties[19].

Greater certainty in the measurement of one property means less certainty in that of another. For example, to measure precisely the *position* of an electron we must view a narrow section of its wave-like nature. But that makes it difficult to calculate its precise *speed* since that requires counting how many waves have passed a given point in a given period of time.[20]

Our perception of reality is at best blurred. If the position and speed of particles can't be known precisely it undermines the idea of a '**deterministic**' universe in which laws of motion precisely define the trajectory of all particles from the past and into the future.

Where's the proof? Fickle nature

Is Nature this unpredictable? Consider a lump of inherently unstable radioactive material. It comprises millions of atoms which decay to stable atoms at a predictable rate. But the precise moment when any single atom will decay is unpredictable. Whilst the decay of any one atom is random, the inherent probability of decay leads to statistically predictable behaviour in a collection of atoms.

This is similar to our earlier coin analogy on page 72. While the result of a single coin toss is unpredictable, on average we expect that 100 coin tosses will yield 50 heads. Radioactivity, it turns out, is Nature's demonstration that it is indeed inherently random.

And in addition, of course, many varied experiments such as the double-slit experiment lend further support to quantum theory.

---

[19] This isn't due to limitations in measuring equipment. Nor to the '**observer effect**' in which the act of making measurements disturbs the thing being measured.
[20] There are other pairs of properties which cannot both be known precisely for similar reasons. E.g. The energy of a quantum state and its duration in time.

THE FIGHT FOR THE NATURE OF REALITY
Many possible interpretations of reality
Quantum mechanics indicates that the electron's choice to be wave-like or particle-like depends on interaction with its environment.

'Reality is in the observations, not in the electron,'

according to physicist, Paul Davies. This indeterminacy left physicists very uncomfortable. The universe had previously been regarded as deterministic, in which particles guided unequivocally by the laws of physics reached predictable futures. Could our sense of reality be dependent on probabilities and wavefunction collapse? The question of if and how wavefunction collapse takes place is known as the **'measurement problem'**. Given that the maths of the wavefunction equation itself works perfectly, the search for a meaningful physical interpretation continues unabated.

Some suggest that, while the maths works, maybe it is 'incomplete', that is: missing some **'Hidden Variables'** which when found would remove the unpredictability of a particle's location.

Another idea, the **Copenhagen Interpretation**, suggests that wavefunction collapse is real; that Nature is inherently unpredictable at this small scale. This does however raise another awkward question. Precisely what does or does not constitute an interaction which can trigger this collapse?

Another interpretation avoids the need for the collapse of wavefunctions altogether. It argues that instead of collapsing into just one state, each particle continues into all possible future states. In this **'Many Worlds Interpretation'** (MWI) a number of parallel worlds come into existence, each invisible to all others. When proposed in 1957, MWI was largely ignored. Now, in straw polls, roughly half of physicists find MWI has at least some value.

There are other interpretations, mostly versions of the above. Whilst the idea of collapsing wavefunctions may not be the ultimate winner, for now it works very well indeed. It explains experimental outcomes, it explains how electronics works, and it also explains how the universe's energy and matter are organised, as we'll see shortly.

Is seeing far more than just believing?
As with relativity, the different interpretations of wavefunctions open up a debate over the nature of reality. Many questions ensue. Is particle interaction sufficient to trigger collapse or is 'observation' required? If observation is required, as some believe, what takes place between the 'observer' and the 'thing observed'? And, is the observer's purposeful intent required, that is, consciousness?

Wanted, dead *and* alive, Schrödinger's Cat

Erwin Schrödinger (who came up with the wavefunction idea in the first place) was unhappy with the Copenhagen Interpretation's suggestion that wavefunction collapse was dependent upon an observer.

By way of demonstration, he invented a hypothetical cat enclosed in a box which also contained a randomly-timed means of causing its death. According to this interpretation of quantum mechanics, he ridiculed, the cat is supposedly *both alive and dead* at the same time – until we open the box to check on its health. At which point the wavefunction apparently collapses and we see either a living or dead cat.

Einstein was as disturbed as Schrödinger by the Copenhagen Interpretation. He added to the cat's troubles by asking if a conscious cat might be aware that it was both dead and alive, or whether the cat's own self-awareness might trigger its own wavefunction collapse! The idea of an unpredictable universe appalled Einstein. Referring to God, he said, '*I am convinced He does not play dice with the universe.*' However, experiments reveal that Schrödinger and Einstein were wrong and that the quantum world is indeed inherently random. (The problem of whether conscious observers are required is highlighted in another intriguing thought experiment known as **Wigner's Friend**.)

We're left with the implication that our universe is not deterministic. It is at best *statistically* predictable. The Many Worlds Interpretation sidesteps this problem by allowing many future states, not just the one chosen seemingly at random. Of course, the idea of many parallel worlds challenges our sense of reality just as much.

However, until there is proof either way we should at least bear in mind that it was only in the last 100 years that we discovered that ours is not the only galaxy, and so likewise ours may not be the only universe!

THE NATURE OF MATTER
Why we don't fall through the floor
If what we imagine to be solid particles are little more than waves of probability which describe how and where their energy is distributed, how come they feel so solid? Add to that that when in a particle-like state these particles take up less than 0.1% of the volume of the atoms they make up. Given this seeming lack of substance, why don't we just fall through the floor? Just how is energy organised to create solid matter?

The electrons orbiting beyond an atomic nucleus are in an energetic state. As such they are inclined to stay in orbit despite their negative charge which would otherwise draw them towards the positively charged nucleus. Furthermore, since they are all negatively charged the electrons keep their distance from each other.

electrons in orbits of whole wavelengths

atom's nucleus

Meanwhile, the wave-like nature of their wavefunctions constrains their vibrations to specific orbits: ones which allow only completely joined up waves, as in the diagram. This is the result of the same constructive interference we saw on page 74. Orbits which don't form complete waves, interfere destructively, and cancel out. (This is similar to the 'stationary waves' observed in a whipped skipping rope.)

It is this energy field of electrons spread out in their orbits which gives substance to atoms. Much as a force-field provides a defensive shield to a sci-fi space ship. It's why we don't fall through the floor.

Why don't mountains appear and disappear?
In everyday life we don't see large objects behaving as if they could be in many places at the same time. So how do the tiny particles which can be in many places at the same time assemble themselves into the large scale predictable reality we observe?

It isn't that the statistical laws of quantum mechanics no longer operate at macroscopic scales, rather that their random effects average out at large scales. Even small amounts of matter comprise a vast number of particles. Since every particle has its own individual wavefunction, each will do its own random thing. Collectively the statistically random behaviour of all the particles averages out. The only way a large object could appear to be in multiple states at the same time would be for the random behaviour of all the particles to be identical.

In the absence of such '**coherence**', as it is called, a large collection of particles takes on the predictable characteristics observed of bigger objects. This is known as the '**Correspondence Principle**'. It states that the quantum mechanical laws of physics must produce the same outcomes as the laws of large scale physics when applied to objects comprising very many particles.

THE SEARCH FOR A THEORY OF EVERYTHING CONTINUES
We're now in a position to see if and how relativity and quantum mechanics can join into a unified Theory of Everything. The short answer is, not yet. Quantum mechanics has been joined with Special Relativity, primarily the result of work by Paul Dirac, widely regarded as second only to Einstein in his contribution to 20th century physics.

> Dirac has done more than anyone this century, with the exception of Einstein, to advance physics and change our picture of the universe.
>
> *Stephen Hawking, Physicist*

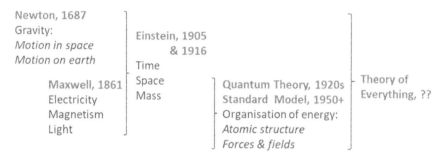

A quantum theory of gravity, of General Relativity, is proving more elusive. Quantum Mechanics is a highly successful theory. Its so-called '**Standard Model**' describes the properties of the universe's most basic building blocks. (Similar to the way the Periodic Table does for the elements.) The Standard Model comprises two types of particle: 'matter particles' called '**fermions**' which make up protons, neutrons and electrons; and, 'force particles', called '**bosons**', which carry the universe's **fundamental forces** from one place to another. [21] (See also page 51.) These fundamental forces are:

- The **strong nuclear force** which binds sub-atomic particles together in an atom's nucleus. It's the strongest force by far but has exceptionally small range.

- The **weak nuclear force,** responsible for the radioactive decay of atoms. It too is stronger than gravity and has very short range.

- The **electromagnetic force**, responsible for electromagnetic waves such as light. Stronger than gravity, it has infinite range.

- The **gravitational force**. It is by far the weakest force, but operates over vast distances.

---

[21] Bosons have been found for three of these forces. One, for instance, is the photon which carries the electromagnetic force. However, as yet there is no quantum description of gravity or gravitational fields. Although a hypothetical gravity boson has already been given the name '**graviton**', there is no certainty that a gravity boson is the way relativity and quantum mechanics will be unified.

Bosons are particle-like fluctuations in their own underlying universe-wide '**fields**'. One example, photons, carry electromagnetic force across the electromagnetic field. Another is the **Higgs boson** and its associated field which, depending on the strength of interaction, imparts an amount of mass to other particles.

However, even an empty field never has zero energy. While, on average, it might be said to have zero energy, quantum uncertainties mean that it is constantly fluctuating, bubbling with energy, causing particles to pop in and out of existence at any moment, albeit briefly.

There are significant differences between relativity and quantum mechanics, hampering the search for a unified theory. Maybe one or both are flawed, maybe new sub-particles or forces will be found, or maybe they're manifestations of another as yet undiscovered theory.

Some of the differences between relativity and quantum mechanics are shown in the table below. The items in bold most closely match our intuitive ideas. It is also clear that they each treat time very differently. This highlights the '**Problem of Time**': that is, of finding an explanation of time consistent with both theories. This is discussed further in the chapter, *On 'Time'*.

Relativity's equations	Quantum mechanics' equations
Space, time & energy have no minimum value. They are 'smooth' quantities.	Space, time & energy have minimum values.[22] They are grainy i.e. quanta.
Time is a part of the universe. Time varies with speed & gravity.	**Time is a separate entity.** **Time is absolute, i.e. constant rhythm.**
Time flows in one direction: **cause & effect rules events.**	Time-symmetry rules: quantum events can run backwards.[23]
Events are (pre)determined by laws.	Events are determined by probability.
The universe is 'local': reference frames cannot be universe-wide.[24]	**The universe is 'non-local':** **distant events can be entangled.**[25]

---

[22] Minimum *measurable* quantities, because of uncertainty, known as '**Planck**' units.
[23] This has been demonstrated experimentally.
[24] We're all in our own 'local' frames of reference. We observe differences in time depending on relative speed. It is not the same 'time' throughout the universe.
[25] Quantum entanglement, see page 83.

Two particles can be **'entangled'** whereby a change of state of one results in an *instantaneous* change of state of the other, even if they are far apart. (Most physicists agree this doesn't infringe the speed limit 'c' since no information is transmitted.)

This synchronised change in the state of particles has been shown experimentally, providing further proof of quantum mechanics.

Experiments also show that entanglement, like energy density, causes spacetime to curve. Is this a clue for how General Relativity and quantum mechanics may be combined?

The search for a Theory of Everything is in effect a search for a quantum explanation of gravity. Central to the challenge are: the smooth / local nature of relativity; and the grainy / non-local nature of quantum mechanics. There are currently two main competing approaches. The first, **'string theory'**, attempts to unify all of physics proposing that particles are actually tiny one-dimensional strings of energy; where different particles are the result of a different vibration of these strings. The second, **'loop quantum gravity'**, attempts to find a quantum explanation of gravity; it envisages tiny volumes of spacetime as the fine quantum grain of the universe.

The search for a Theory of Everything takes physicists to the frontier of physics. Gravity, the weakest force, has a tiny effect on particles of small mass. However, in massively dense black holes the force of gravity is of comparable scale to the other forces. Here, maybe, a quantum explanation of gravity can be discovered.

IN THE BEGINNING, SOMETHING FROM NOTHING

On page 60 we described how the universe exploded into existence. This 'Big Bang' happened 13.8 billion years ago. (The earth is only 4.5 billion years old and Homo sapiens appeared less than 1 million years ago.) Random fluctuations in energy fields in apparently 'empty space' snow-balled almost instantaneously, an event known as **'inflation'**. A vast quantity of energy was released and organised itself into the energy, matter, forces and fields we see today, as the diagram on the next page shows.

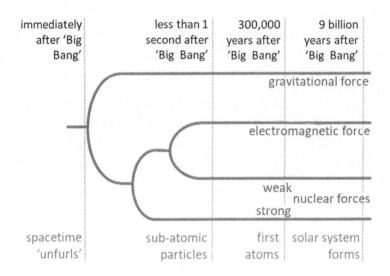

immediately after 'Big Bang'	less than 1 second after 'Big Bang'	300,000 years after 'Big Bang'	9 billion years after 'Big Bang'

gravitational force

electromagnetic force

weak
strong — nuclear forces

spacetime 'unfurls'	sub-atomic particles	first atoms	solar system forms

> Everything – space, time, energy and matter – came into being in the Big Bang and began expanding everywhere at once.
> *Marcus Chown, Science writer*

Mistaken Reality, Again

Once again, our evolution-constrained senses have misled our ideas of reality. If we had everyday experience of the quantum world we would think particles behaving unpredictably was perfectly normal.

If this brief overview of quantum weirdness has left you a little enlightened and a lot bemused, you are in good company.

> Anyone who says that they understand quantum mechanics does not understand quantum mechanics.
> *Richard Feynman, Quantum physicist*

Only time will tell if we'll ever be able to peer beneath Nature's veil of quantum unpredictability. Indeed, maybe 'time' has the answer.

Chapter

On 'Time'

Time, Not What it Seems

> Put your hand on a hot stove for a minute, and it seems like an hour. Sit with a pretty girl for an hour, and it seems like a minute. That's relativity.
>
> *Albert Einstein*

Einstein was clearly sharing a joke, but one with a point, highlighting as it does our variable perception of time. He invites us to question that perception; to stand not in the rushing waters of time's river but to view it from above, to try and comprehend the whole river. This chapter describes the current state of thinking about 'time'.

The so-called Problem of Time was summarised on page 82. Not only do relativity and quantum mechanics not agree with each other about the nature of time but they are also at odds with our intuitive perceptions of time.

There is a historical irony to this. In 1905, when relativity first demolished our intuitive sense of time it had been barely twenty years since countries first started synchronising clocks across their territories. This had been done to assist train timetabling and signalling. Before that there was no earth-bound 'right time', nor any suggestion that the very concept of time was flawed. However, it is worth noting that our intuitive idea of time has obvious contradictions of its own.

The problem with infinite time

In his 1781 *Critique of Pure Reason*, philosopher Immanuel Kant described the implicit contradictions of time stretching infinitely into the past or future. If the universe had been created, he said, what made it wait an infinite time before coming into existence? On the other hand if it has been around forever everything that could happen will have happened and the universe would have long ago ceased to exist.

More Problems of Time. Past, Present or Future?

There are other Problems of Time. Most obviously, we've never seen time. This may seem an absurd statement, but given how relativity has changed our notion of space and time, it is worth noting.

One problem comes from the Relativity of Simultaneity. We saw on page 31 that observers moving relative to each other disagree about which of two events occurred first. They disagree about what 'now' even means. It's possible that one observer might have witnessed event A and not yet be aware of event B, whilst the other may have witnessed event B and not yet be aware of event A. In other words they are disagreeing about what lies in the past and what lies in the future! (This is sometimes known as the '**Andromeda Paradox**'.)

This suggests that all events have already taken place whether or not we think they have, otherwise how else can we explain the above paradox? This idea, known as the '**block universe**', is troubling. Does the future already exist? To us humans a distinction exists between past and future: the past leaves traces, the future does not. Whilst this distinction feels real we must ask ourselves, "Is it actually real?"

The block universe idea gains support when we consider that space and time were created together in the Big Bang. Since we perceive 'space' as all around us ('here' and 'there') then all of 'time' might also be all around us ('now' and 'then').

> People like us, who believe in physics, know that the distinction between past, present and future is only a stubbornly persistent illusion.
> *Albert Einstein, in a condolence letter to the widow of his life-long friend, and just a few weeks before his own death*

We are now at the very limits of physicists' knowledge and no one quite knows if we really are in a block universe or something that mimics a block universe. What is clear is that, since light-speed is finite, the concept of 'now' (the present) only makes sense for events taking place in close proximity to us individually. By the time we witness anything from farther afield, it actually occurred at some time in our own past.

PERSPECTIVES ON TIME

We are left with many questions, the most obvious of which, 'What is time?' still troubles physicists and philosophers as it has for millennia.

Other questions involve the flow of time. If we perceive time as flowing from the future toward the past does that mean that time actually flows; and if not, why does it *appear* to flow (while space just 'is')? And why does it flow in only one direction, compared to space through which we can travel in all directions? Then there are questions about the 'present'. What is the 'present'? Why is it that we are only conscious in what we perceive as the 'present'?

Physicists don't have answers to these questions, but there are some common ideas as to where the answers might lie.

Aside from the many differences between relativity and quantum mechanics discussed on page 82, there's one word in the table which appears in both columns: both theories recognise the significance of *events*. Physical processes *flow* from one event to another. And these processes are encapsulated in the laws and equations which describe that *change*.

Let's take a closer look now at some current ideas about *events, flow,* and *change* and their relevance to the Problem of Time.

Events they are a changin' (With apologies to Bob Dylan)

> The only reason for time is so that everything doesn't happen at once.
> *Albert Einstein*

(We might more fully appreciate the idea behind this quote by considering its corollary: the only reason for space is so that everything doesn't happen in the same place!) Put another way, if nothing ever changed around us, if there were no events, then there would be no moment-to-moment differences to observe. Arguably, without these moment-to-moment differences there would be no concept of 'moments' themselves, and so no need for time. Things that we might see in a permanent state would just be events which, for a long while at least, are not changing.

Put another way, does time have an existence independent of events? Maybe, the universe is 'just' a network of events governed by quantum mechanical processes. New events are the result of random quantum interactions between past events. This suggests that there is no inherent order; rather, order emerges as interactions occur and time is the means by which we sub-consciously order or count how events change. (This is an idea first put forward by Aristotle nearly 2,500 years ago. It is also reminiscent of one of Zeno's Paradoxes, around the same period, challenging our idea of motion by asking whether an arrow in flight is actually moving at any given instant.)

If this is the case, then time has no separate or real existence. When time appears in the equations of physics it isn't describing a real thing, such as distance, but is a reflection of some underlying process of the universe. If so our human sense of time would be a construct of our minds: the memories of past events or the anticipation of future events. Events pieced together like a movie comprising many individual frames, giving the impression that there is such a thing as time and that it flows from one frame to the next.

> Time has no independent existence apart from the order of events by which we measure it.
> *Albert Einstein*

The one-way flow of time
Why we perceive this flow, real or otherwise, in only one direction, past to future, baffled physicists until the mid-19th century because no known law of physics requires time to be directional. Consider the simple equation:

SPEED equals DISTANCE travelled divided by TIME taken.

Nothing about this equation restricts time to only one direction.

The one way flow of time seems all the more surprising when we consider that, as mentioned earlier, quantum processes themselves can flow backwards, and do, as has been experimentally observed.

If the laws of physics don't restrict time to one direction, why do we perceive time as irreversible? The answer has nothing to do with relativity or quantum mechanics and everything to do with energy and probability.

A tidy pile of bricks can be easily knocked over. As it crashes to the ground energy is lost randomly to noise and heat. Conversely, we can't imagine a load of bricks, lying randomly on the ground, assembling themselves just as easily into a pile. There are so many more ways for the bricks to be in an untidy mess than in a nice neat pile. It takes concerted effort and energy to rebuild the pile as it was. Of course that's not impossible just highly improbable without that effort and energy. (We feel the same when asked to tidy a room. It was easier to leave a mess, it takes greater effort to tidy it up.)

Likewise, what freely moving vibrating atoms actually do is collide randomly with other atoms. As they do some energy is transferred between them and some is dissipated, for example as heat. As a result, over time, the statistically likely outcome is that all objects and atoms eventually reach energy equilibrium with each other.[26]

This tendency towards disorganised states is called '**entropy**' and imposes itself across the universe. It is why we observe events unfolding in only one direction. 'Past' and 'future' are distinguished by this statistically more likely transfer of energy. This so-called '**arrow of time**' is accepted physics. Beyond that is speculation.

---

[26] The vibration of atoms is what we experience as heat. The part of physics, which relates heat to other types of energy is called **Thermodynamics.**

Some believe that time's arrow might be linked to the expansion of the universe: as it expands it thins and cools, its energy gradually dissipated evenly throughout. Without any energy differences providing the motivating means to do work, events cannot happen and order cannot be created; our universe is like a heart monitor gradually flat-lining.

Some physicists go further, suggesting that the expansion of the universe is linked to our perception that time flows. So, whilst we see the dimensions of space around us, the perception of a time dimension rushing past is the result of space expanding.

In Conclusion

A number of things seem clear.

Firstly, that we are far from an answer to the question, 'What is time?'

Secondly, that time is almost certainly nothing like our intuition would have us believe; indeed it may not even exist as an entity in and of itself.

And thirdly, that the Problem or Time lies at the frontier of physics and may provide the clues that lead toward an eventual Theory of Everything.

Chapter
RELATIVITY'S PHILOSOPHICAL IMPLICATIONS

This chapter touches upon some of the philosophical implications of relativity and quantum mechanics, where the cutting edge of physics meets the timeless discussion of philosophers. These ground-breaking advances in 20th century physics shine new light on age-old questions although they still don't come up with complete answers.

THE LIMITS OF PERCEPTION, HOW MUCH DO WE REALLY KNOW?

The Alice in Wonderland nature of length contraction, slowing time, and the surprises of spacetime force us to confront the fact that our everyday experience is confined to less than one millionth the speed of light. Quantum mechanics further challenges our insufficient macroscopic senses with the random nature of the sub-atomic world.

> Our best theories are not only truer than common sense, they make more sense than common sense.
> *David Deutsch, Physicist*

That our senses so mislead forces the realisation that they have been constrained by evolution for survival in a low-speed, large-scale world. If our evolution-constrained senses lead us to be so unaware, we must forever be conscious not just of our limitations, but that we may not even know what those limitations are.

To disbelieve just because it is not what we 'expect' is pure arrogance – a trait we humans are prone to demonstrate. Modern physics highlights just how little we actually know, despite how well we think we know it!

This dependency on our unreliable senses brings to mind Plato's allegory of the cave inside which people are forever captive. In this allegory, they can see only the shadows of reality which are cast onto the cave's walls. And so these shadows become the cave-dwellers' reality, a *perceived* reality. We are, in similar fashion, prisoners of our somewhat deficient senses.

How Real is Reality?

Given these sensory deficiencies it seems reasonable to ask, just what is 'reality'? Added weight is given to this question by the Uncertainty Principle's implication that there is no such thing as absolute certainly. Quantum mechanics further demonstrates that observer and thing observed are interdependent. Some suggest that reality exists *only* when observed. Some people ask the further question, does the thing doing the observing need to be conscious? A version of the well-known question, 'If a tree falls in a forest and no one is around to hear it, does it make a sound?'

I Do Maths, Therefore I Am

There are those who would take this logic one step further. If our description of the universe through relativity and quantum mechanics is so reliant on complex maths, then is *what* we perceive dependent in some way on the form of mathematics we've invented?

> The most incomprehensible thing about the world is that it is comprehensible.
> *Albert Einstein*

This leads to an uncomfortable choice. Would we perceive the universe differently if we invented a different form of maths? It feels safest to presume this can't be right. After all, the maths we have invented works so well at explaining our world experience. In which case mathematics must in some way be fundamental not just to a description of the universe, but to the universe itself! This would mean that we did not invent maths, we merely discovered it. In 1960, physicist Eugene Wigner described this as *The Unreasonable Effectiveness of Mathematics in the Natural Sciences.*

But if mathematics is fundamental to the design of the universe, would it not have had to exist at the time of the universe's creation? This presumably includes the creation of the maths and physics needed to design the universe in the first place. For now all that can be said with any certainty is that our laws of physics, as they currently stand, do not work at singularities such as the Big Bang itself.

Quite possibly, much as evolution theory explains how complex organisms arise from simpler ones, so the many complex laws of maths and physics evolved from far simpler origins.

We might also ask if we and all we experience are just the result of complex calculations by an enormous computer we cannot perceive. And if so, could we ever know it? Much as a computer game character presumably does not know that it is a virtual creation.

> In the beginning the Universe was created. This has made a lot of people very angry and been widely regarded as a bad move.
> *A Hitchhiker's Guide to the Galaxy, Douglas Adams*

THE BIG BANG, CHICKEN AND EGG

Some models of the origin of the universe suggest that in the miniscule fractions of a second after the Big Bang time was exactly like a dimension of space, but very soon after it unfurled into the dimension we now experience as time.

'What came before the Big Bang?' is not an easy question to answer since, if there was no time, there can be no 'before'. Physicist Stephen Hawking explains that that's a bit like asking, 'What's north of the North Pole?' Equally, 'Where was that Big Bang?' or 'What is spacetime expanding into?' are not easy questions either: the answers 'nowhere' and 'nothing' are insufficiently satisfying.

CAN EMPTY SPACE BE EMPTY?

This question was first posed by ancient Greek philosophers as a conundrum. They argued that if two objects are separated by empty space, by 'nothing', then there is 'nothing' between the two objects, therefore they must be touching!

Quantum mechanics reveals that maybe the Greeks were onto something. As we've seen, empty space without any electrical or magnetic charges and without any matter at all is nonetheless permeated with fields; and that from this 'nothing' came the 'something' that makes up the universe. Here, once again, we meet the limits not just of the universe, but of our current understanding.

THE DESIGN OF THE UNIVERSE

In the Big Bang theory the universe can bring itself into existence by virtue of the laws of physics. Once again, we must ask, where did those laws come from? Did scientists *discover* them, or did they *invent* a version that works at explaining the universe?

The conundrums don't end there. As we've seen, the speed of light 'c' plays a fundamental role in the structure of the universe. This universal speed limit, one of nature's constants, establishes the relationship between space and time, and between energy and matter. But the existence of this constant and others like it yet again begs the question: where and when did they arise?

To highlight the concern, physicist Arthur Eddington asks us to imagine a sculptor's claim that the form of a human head lies within a block of stone. Incredulous, we watch as the sculptor hammers and reveals. Was it just the result of skilful chiselling, or was *this* head actually there before?

What if the sculptor had made the nose longer? What if the finely-tuned relationships between the universe's various constants were just marginally different? Might the universe and all the life in it be unrecognisably different, or indeed not have come into existence at all? This is known as the '**Anthropic Principle**'. In its extreme form it suggests that the universe exists only to bring us, human observers of the universe, into existence! This seems a very self-centred interpretation, reminiscent of the belief held for millennia that the earth is the centre of the heavens, and that Man is the purpose of all Creation.

If we do have a special place in the universe how are we to make sense of that? To some, the Many Worlds Interpretation of quantum mechanics provides a clue. If there are many parallel universes, all invisible to each other, Mankind will come into being only in the tiny subset of these universes in which conditions are favourable. And, unable to see the other parallel universes, Mankind will consider himself special. Much as a toddler who has not yet developed a 'Theory of Mind' perceives only its own importance.

Free Will?

Newton's universe was deterministic, the laws of physics guiding particles along inevitable paths. Tangential support for this comes from relativity, specifically the block universe in which the future as well as the past may already exist (see page 86).

Meanwhile quantum mechanics seems to tear up any notion of determinism in favour of indeterminable random outcomes. Perhaps, after all, our instinctive belief in free will is somehow proof that the universe must be indeterminable? Or is that simply another illusion yet again masking what actually goes on behind the curtain of reality?

Can the Universe Actually Be Deterministic?

One of the obstacles to finding a Theory of Everything appears to be that the possibly predetermined view of relativity is at odds with the inherent uncertainty of quantum mechanics. But can the precise position or speed of objects be truly known even in classical physics?

Working out particle trajectories, of necessity, involves some numbers with an infinite number of decimal places: calculations such as 3/2, or constants such as π (pi). As such, future states can't be known precisely because the numbers themselves aren't precise numbers. (This is known as '**intuitionist maths**'). The chaotic knock-on effect of these now imprecise calculations seems to have more in common with the less than knowable quantum world, than with the presumed precision of deterministic models of the universe.

Pure Chance?

One way or another probability is prevalent in the operation of the universe. The laws of physics that govern its operation seem inherently rooted in chance. Be that the statistical uncertainty of quantum mechanics (pages 75-76); or, the statistically directed arrow of time (pages 89-90); or, the fortunately anthropic alignment of universal constants which gave rise to *our* universe (page 94). The very universe in which we primates lead our inquisitive search for how the universe in which we ask these questions came to be.

> We are just an advanced breed of monkeys on a minor planet of a very average star. But we can understand the Universe. That makes us something very special.
>
> *Stephen Hawking, Physicist*

Pure chance seems to me to be a perfectly acceptable explanation for the universe and its design. After all a large enough number of monkeys will eventually type out that design, because, well, that's the sort of thing a vast number of monkeys with typewriters do.

ALL MIGHTY?

There are those who invoke a God-designer. They ask, "How else could all this improbability come together to form a coherent universe?" Such ideas seem unsatisfying, leaving answered the question, 'But where did She or He come from?'

When we consider the laws of physics we describe them as: 'universal', 'absolute' and 'inviolable'. These are similar terms which (some) humans use to describe an omnipotent God. And, let's not forget that an omnipotent God is also often seen as being at odds with humankind's sense of 'free will'.

Relativity and quantum mechanics point to a different description of the universe than that proposed by religion and its ideas of God. Given the light they shed on its design, we might imagine that if there is a God then He or She is almost certainly a mathematician!

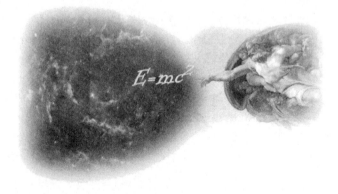

Chapter
RELATIVITY, CULTURE AND SOCIETY

This chapter takes a very brief look at some of relativity's wider influence on human thought.

SPARKING IMAGINATIONS

Relativity was largely unknown outside the scientific community until the 1919 experiment that showed starlight could be bent by gravity. In confirming Einstein's theory that space and time were not absolute it shook the very foundations of beliefs Mankind had held since time immemorial. As such, it made front page news worldwide. And it seeped into the imaginations of thinkers and creative spirits across diverse fields.

IDEOLOGY

We have already seen that proponents of Relativism wrongly deployed relativity as proof of their point of view that nothing at all was absolute (see page 61). But others also employed relativity to support their ideas.

There were those who interpreted the almost purposeful laws and constants of physics as a demonstration of the universe's divine spirituality.

At the other extreme, some Soviet physicists argued that relativity's emphasis on reality as it really exists supported a Marxist materialist agenda, rather than the idealised sense of reality which we as humans perceive.

And 'logical positivists', who believed that logic applied to empirical evidence was the sole means by which to arrive at truth, deployed relativity in support of their views.

Relativity was a bandwagon on to which many people climbed. By and large it seems that Einstein did not appreciate all these fellow travellers.

> I am not a Positivist. Positivism states that what cannot be observed does not exist. This conception is scientifically indefensible, for it is impossible to make valid affirmations of what people 'can' or 'cannot' observe. One would have to say 'only what we observe exists,' which Is obviously false.
>
> *Albert Einstein*

The worst ideological abuse was the attack on relativity on the pseudo-scientific grounds that it was manipulative 'Jewish Physics'. This occurred primarily in Germany in the years between World Wars I and II. Einstein and his relativity were attacked, as was quantum mechanics, the development of which involved some Jewish physicists. The anti-Semitic narrative branded the new physics as antithetical to the principles and supremacy of Aryan beliefs. A perfect storm of factors gave this narrative momentum: the petty jealousies of a high profile German scientist, the rising tide of populist opinion, and the Nazi regime's adoption of this narrative in furtherance of its propaganda and abhorrent actions.

ART AND LITERATURE

The 'Modernist' movement had been accelerating from the 1850s, rejecting the certainty of earlier Enlightenment thinking. It appealed to innovators to 'Make it new!' in all fields, lasting well into the 1900s.

Relativity's later arrival is of course not a response to that appeal. But strong arguments can be made for the influence relativity had on Modernists. The case here is not that relativity added momentum to Modernism, but that the uncertainties of space and time provided a theme which could be used in art and literature.

As such relativity influenced many styles such as Dadaism and Surrealism. *The Disintegration of the Persistence of Memory* by Salvador Dalí is one example widely cited in which the relationship between mass, energy, space and time were explored, though Dalí himself denied this influence. MC Escher's work, *Relativity,* is another example, depicting the complete dislocation of space.

Relativity
by MC Escher

The equation $E=mc^2$ has become widely used as: a logo to motivate staff to work smarter; a dramatic plot device; song titles; and art installations; to name just a few. Anish Kapoor has explored humans' psychological disassociation with space and time caused by extreme darkness, by placing people inside a completely matt black, dark space. Any notion of space eventually disappears as they lose all sense of physical reference points or objects. Since time is perceived as the progression of events, they lose any sense of time also.

The subjective nature of time was also taken up by writers.

> If all time is eternally present
> All time is unredeemable.
> What might have been is an abstraction
> Remaining a perpetual possibility
> Only in a world of speculation.
> *Four Quartets, T.S. Eliot*

And, in *2001 A Space Odyssey* we follow one of the film's unwitting astronauts through a 'Stargate', a singularity in spacetime.

> In a moment of time, too short to be measured, space turned and twisted upon itself ... Not only space ... The seconds themselves were passing with incredible slowness, as if time itself was coming to a stop.
> *2001 A Space Odyssey, Arthur C Clarke*

In another example, Kurt Vonnegut gave a capacity for time-travel to an alien species as a means of understanding the character of his all-to-human protagonist.

> All moments, past, present and future, always have existed, always will exist. The Tralfamadorians can look at all the different moments just that way we can look at a stretch of the Rocky Mountains, for instance. They can see how permanent all the moments are, and they can look at any moment that interests them. It is just an illusion we have here on earth that one moment follows another one, like beads on a string, and that once a moment is gone it is gone forever.
>
> *Slaughterhouse 5, Kurt Vonnegut*

GOING NUCLEAR

Mass-energy equivalence, as described by $E=mc^2$, pointed to the new horizon of the nuclear age. Quantum mechanics provided the tools which enabled engineers to realise that potential. Within a few decades it gave society a whole new set of moral dilemmas such as: atomic weaponry versus 'mutually assured destruction', and nuclear energy versus fossil fuels.

The ensuing debates seem to pit one person's moral high ground against another person's real world pragmatism, with no middle ground upon which either side feels able to compromise or agree on a science-based balanced risk assessment.

Definition
going nuclear
Taking things to the absolute extreme in order to avoid a series of small escalations. This can be a way of winning a fight you might not otherwise win, but has the potential to destroy both people involved.
(From urbandictionary.com)

A REMARKABLE 150 YEARS

The 150 years from 1850 to 2000 saw a revolutionary advance in science. Presuming this exponential rate of discovery continues we can only imagine what the next 150 years will bring, and how that in turn will influence the world beyond physics.

Chapter
MORE ABOUT EINSTEIN

Einstein was:

> '*unfathomably profound — the genius among geniuses who discovered, merely by thinking about it, that the universe was not as it seemed.*'

This is how TIME magazine explained why Einstein was their choice for 'Person of the 20th Century'. They continued that this choice was part due to the '*sheer brilliance*' of his work on relativity, and part due to the work's '*far reaching implications*'.

It seems fitting that a magazine called TIME named Einstein as its 'Person of the 20th Century' given relativity's insights into the nature of time itself.

WHAT MAKES EINSTEIN SO CLEVER? CREDIT WHERE CREDIT IS DUE

By 1905, when Einstein published his two papers on Special Relativity, a few scientists had thought of many of the things that Einstein thought and had created much of the maths that he would need. Indeed, in 1953 Einstein himself reflected that relativity was '*ripe for discovery in 1905*'. Notable work on Special Relativity was done by George Fitzgerald, Joseph Larmor, Hendrik Lorentz, Henri Poincaré and Herman Minkowski, as well as Einstein. This has led naturally to some debate as to whether Einstein deserves all the credit he gets.

But these other scientists didn't arrive at Einstein's elegantly consolidated view of the universe. (In one of his 1905 papers on Special Relativity the portion dealing with the mechanics and maths of travel at or near the speed of light was just twelve pages long.)

Others had persisted in trying to make the theory and the maths fit old models, such as the existence of the ether. It is as if the majority of the scientific community treated Michelson-Morley's surprising observation of the invariance of the speed of light as a problem which had, somehow, to be fitted into their view of the universe. Einstein on the other hand started afresh with a blank sheet of paper discovering in the process that the invariance of the speed of light, far from being a problem, was the key to a door behind which lay a whole new universe.

> Talent hits a target no one else can hit.
> Genius hits a target no one else can see.
> *Arthur Schopenhauer, Philosopher*

Furthermore, the General Theory of Relativity was undisputedly individual brilliance. And history's jury is in little doubt that Einstein deserves the credit for relativity. Hendrik Lorentz, himself significant in relativity's early development, put it as follows:

> *'I considered my [work] only as a heuristic working hypothesis. So the Theory of Relativity is really solely Einstein's work. And there can be no doubt that he would have conceived it even if the work of all his predecessors in the theory of this field had not been done at all. His work is in this respect independent of the previous theories.'*

Hope for us all?

Einstein's younger sister found the young Albert a bit dreamy and slow. His parents, according to Einstein himself, *'were worried because I started to talk comparatively late, and they consulted a doctor because of it.'* As a child he was fascinated by a compass he'd received as a gift and that the needle could be made to move by something unseen.

The myth that he did poorly at school is, just that, a myth. In fact, throughout his school days he did very well, rebelling only against the 'by rote' learning; an early sign, perhaps, of his questioning nature.

He was initially declined a place at university which is what led to him taking a job at the patent office in Bern. There, amongst other things, he reviewed patents for machines that, rather ironically in hindsight, enabled clock synchronisation for the burgeoning railways.

1905, NOT A BAD YEAR

Einstein published four important papers in 1905, his so-called **'Annus Mirabilis'**. Any one of these were worthy of a Nobel Prize. The first on photo-electricity proposed the idea of energy quanta, see page 73. The second, on Brownian Motion, proved the validity of the, until then, theoretical atomic model of matter. The last two were his papers on special relativity: one on length contracting and time slowing, and one on mass-energy equivalence.

BUT HE WASN'T ALWAYS RIGHT

Einstein deserves his elevated place in the history of ideas. But no one is infallible. Here are a few things he didn't get right.

- Einstein resisted his own equations' conclusion that the universe was expanding, and forced an ad hoc **cosmological constant** into them so that they described a static universe. (See *A Little Bit of Maths*, page 129, for more.) In later years he apparently described this as, *'the biggest blunder'* of his life.

- Despite developing the ideas that lay behind 'spacetime', Einstein initially felt that the **spacetime diagrams** developed by Herman Minkowski added little of value. Later, he saw their immense value and used them in his development of General Relativity. (See *A Little Bit of Maths*, page 109, for more.)

- Einstein at first did not believe that **gravitational waves** were a valid implication of relativity, but he later changed his mind.

- Einstein argued against the idea of spacetime singularities such as those required by the **Big Bang** or found in **black holes**.

- *'He does not play dice,'* referring to God, is one of Einstein's most famous quotes in response to the probabilistic nature of quantum mechanics'. But his intuition has been proved wrong. (See page 78.)

- Einstein wrongly disputed the evidence for 'quantum entanglement' in which two particles know about each other's change of state faster than the speed of light should allow (see page 83). He called it, *'Spooky action at a distance'*.

WITH REGRET

These last two errors of judgement disappointed Einstein's many friends in the physics community who held him in high regard; and who saw him as a founding father of our quantum understanding of the world because of his 1905 paper on photo-electricity and quanta (see page 73).

Einstein had successfully challenged so many previously held theories and perceptions of reality, and yet he could not accept the random nature of quantum mechanics and continued his search for a classical meaning to the new quantum equations.

> Many of us regard this as a tragedy, both for him, as he gropes his way in loneliness, and for us, who miss our leader and standard bearer.
> *Max Born, Physicist*

WHY IS EINSTEIN SO FAMOUS?

Quantum mechanics is arguably as important as relativity. However, no scientist is as singularly and firmly engraved on the public imagination in respect of quantum mechanics as Einstein is with relativity. This was partly due to the fact that quantum mechanics evolved over a longer period of time and resulted from a far more collaborative body of work.

But of course Einstein's discovery that space and time were not absolute, proven by the 1919 demonstration of the bending of starlight, was of more than newsworthy interest, pulling the rug as it did from Mankind's sense of place in a hitherto absolute universe.

> The ideals of Aristotle and Euclid and Newton ... the basis of all our present conceptions prove in fact not to correspond with ... the fabric of the universe.
> *The Times of London, 1919*

The enormity of his discoveries propelled the single name Albert Einstein onto front pages everywhere, making the name 'Einstein' a household proxy for the word 'genius'.

WAR AND PEACE

Whilst not involved in research related to nuclear energy, Einstein's pre-eminence and reputation was a door-opener to political leaders. That's why nuclear physicists, worried about possible Nazi development of a nuclear weapon, enlisted his assistance.

Having escaped Nazism and with certain knowledge of successful experiments in Berlin, Einstein shared their concerns and readily agreed to write to US President Roosevelt in 1939, thereby helping initiate the US's early research into such weapons.

But in the wake of the atomic bombings of Hiroshima and Nagasaki, and in the light of history, Einstein came to regret his *'great mistake'*:

> *'Had I known that the Germans would not succeed in developing an atomic bomb, I would have done nothing.'*

Einstein himself was in actual fact a lifelong pacifist. This included his efforts after World War II to promote the international control of nuclear weapons. Asked to explain how people could discover nuclear weapons but not the means to control them, he replied:

> *'Because politics is more difficult than physics.'*

The irony of Nazi bigotry

Ironically the German failure to build a nuclear weapon was in part a product of the Nazi attack on the new physics of relativity and quantum mechanics discussed earlier (see page 98). It contributed to an exodus of world-renowned physicists prior to World War II from Germany and from other European countries likely to fall under its influence. They left to find work (many having been barred from university posts in Germany), or out of principled objection, or fear; and in search of institutes working on the new physics rather than reviling it.

Their work in those pre-war years paved the way to the harnessing by Man of nuclear reactions. They broke with the academic tradition of collaboration and kept some of their scientific discoveries unpublished and unshared, thereby thwarting German nuclear weapon development.

Eventually, in the midst of war, some of these physicists converged in a remote region of New Mexico, together with thousands of others, part of a secret US project: nurse-maids to the very first nuclear weapons.

IMAGINE EINSTEIN'S EXCITEMENT

Einstein was a 26 year old German-born clerk in the Swiss Patent Office when he published his first paper on relativity in 1905. This, together with his subsequent papers and culminating with his 1916 paper on General Relativity, amounted to a hugely significant advance in the search for a Theory of Everything: that magnetism, electricity, light, motion, matter, energy and gravity amongst other things are all manifestations of spacetime in one form or another.

Probably, as he first tugged at a loose thread exploring the nature of light, he had little idea of the extent to which Newton's universe would unravel; or that it would lead to a vastly different conception of the universe. Did he jump for joy or did he sit still with a sense of awe?

Of course scientific advance is not achieved in such splendid isolation nor in just one evening with paper and pencil. Indeed some of relativity's concepts were emerging before Einstein and some of its implications emerged later through the work of others. But there is no doubt that Einstein experienced a version of this joy and awe.

Something actually snapped

Referring to the moment Einstein proved to himself that General Relativity explained the unusual orbit of Mercury and hence that this theory over which he'd been toiling was correct, Einstein's biographer Abraham Pais wrote, "This discovery was, I believe, by far the strongest emotional experience in Einstein's scientific life, perhaps in all his life. Nature had spoken to him. Whilst the great man himself said,

'For a few days, I was beside myself with joyous excitement.'"

Einstein is reported to have told a friend later that he felt as if something in him had actually snapped.

(From *Subtle is the Lord: The Science and the Life of Albert Einstein*, by A. Pais)

Imagine then how it must have felt to be so instrumental in all these discoveries. And for a brief period to be the only person anywhere who had ever looked this far behind the curtain of reality.

Chapter

A LITTLE BIT OF MATHS (IF WANTED)

This chapter is a wholly optional read for non-serious mathematicians, and uses only high school maths. In ten parts it presents a simplified version of some of the mathematical reasoning which lies behind relativity theory.

Special Relativity
1 Speed limit c. Time slows. Length contracts. Frames of reference
2 Spacetime diagrams. Cause and effect
3 Simultaneity and Causality
4 Explaining paradoxes
5 The Relativity Factor
6 Transformation of coordinates and speed between moving frames
7 Spacetime interval
8 The maths behind increasing mass
9 Arriving at mass-energy equivalence, E=mc^2
General Relativity
10 Einstein's other famous equation: General Relativity

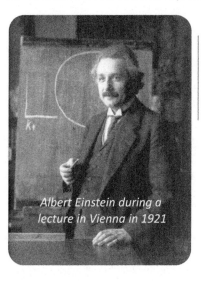

Albert Einstein during a lecture in Vienna in 1921

The equations of general relativity are his best epitaph and memorial. They should last as long as the universe.
Stephen Hawking, Physicist

(Note. In this chapter x, y, z & t are used where, strictly-speaking, Δx, Δy, Δz & Δt should be written. Here Δ (Greek letter 'delta') means 'amount of change in' x, y, z, or t. For clarity we'll omit Δ, taking this as understood.)

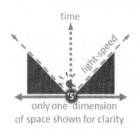

only one dimension
of space shown for clarity

Travelling in time; and The universe's speed limit
A graph of space versus time. Time is shown flowing vertically for the observer ('S'). We can regard her as moving through time. 'S' is 'stationary' in space, i.e. not moving sideways. Light travels at light-speed: a given distance in a given time. Travel faster than light-speed, c (into the black areas) is impossible.

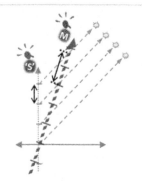

Time slows
A moving observer (M) moves relative to 'S', so his path angles away as he is moving in space. Once a second, as 'S' moves up the time axis, she flashes a light. Of course, these flashes travel to M at light-speed c. Observing their arrival, as they intersect his time axis, M sees 'S's 1-second intervals as longer, i.e. slower, than his own.

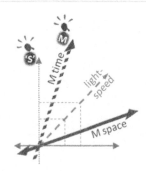

Frames of reference; and Relative motion
The speed of light must be the same for 'S' and M. So M's axes of time and space must be equidistant from the line for light-speed (as they are for 'S'). This way they calculate the same speed (distance/time) for the light beam. As a result 'S' sees M's reference frame transformed, in the way shown on page 45. And the faster the relative speed, the greater the transformation.

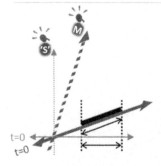

Length contracts
A black rod is moving with M. At t=0, it rests on M's space axis. As seen by 'S', from her frame of reference the rod looks shorter. In each case, the observers are measuring the distance between the rods' two end points *at the same time*, i.e. t=0, *within their own reference frame.*' Since, in relativity, that's how 'length' is defined.

2 SPACETIME DIAGRAMS. CAUSE AND EFFECT

In 1908, mathematician Hermann Minkowski, building on Einstein's work, described spacetime as a single entity, and developed a mathematical tool, '**spacetime diagrams**', to help solve problems in relativity, suggesting that:

> *Henceforth space by itself, and time by itself, are doomed*
> *to fade away into mere shadows, and only a kind of union*
> *of the two will preserve an independent reality.*

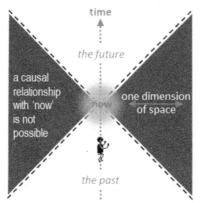

In this spacetime diagram only one dimension of space is shown for clarity. The vertical dotted line represents an observer's journey through time.

Since the observer believes himself 'stationary' in his reference frame there is no movement sideways in space. He travels only upwards in time, along his so-called '**worldline**', towards the future. From his perspective, time is flowing past him. We tend to use the word '**now**' as loosely as the word 'here', hence the fuzzy 'now' in the diagram above. However, the finite speed of light limits what we can actually know is happening 'now' to those events which are literally next to us.

Objects, including light, can move through the observer's reference frame, each with their own worldline. (The diagrams on page 108 are spacetime diagrams which include moving objects.) If their relative speed is non-zero, their worldline will be at an angle to the vertical. The worldline of light beams will always be at 45 degrees to the 'stationary' observer's worldline, as this is the maximum possible light-speed. (45 degrees if the axes are: 1 vertical unit equals 1 second of time; and, 1 horizontal unit equals 1 light-second of length (the distance light travels in 1 second.))

The white region forms a '**light cone**' within which an event taking place *now* could have been **caused** by a past event or could cause a future event. However, it could not be causally related to events in the black region since that would require travel faster than c. (See also pages 110 and 117.)

3 Simultaneity and Causality

Simultaneity

At left, below, we see the reference frame for stationary obserer 'S'. Events A and B are simultaneous in 'S's frame of reference; i.e. they take place at the same time on 'S's wordline, along the same horizontal dashed line. This line is called a **'line of simultaneity'**. These lines of sumultaneity always run parallel to the space axis.

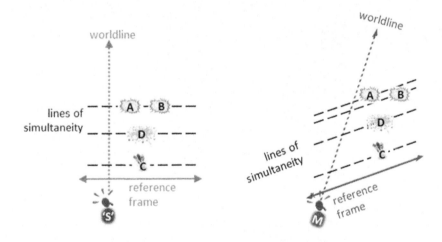

At right above we see the transformed reference frame of a moving observer, M, as seen by 'S'. The lines of simultaneity also rotate with the moving frame's time axis, in order to remain parallel to M's space axis. As a result, from M's perspective, events A and B are not simultaneous since light from these events arrive at different times on his worldline.

Causality

(*See also page 109.*)

However, consider the case of event C (the lighting of fireworks) and event D (fireworks exploding in the sky). Light from event C will arrive before light from event D at both 'S's and M's worldline. In other words, 'cause' will always precede 'effect' no matter how fast M is moving relative to 'S'. This is because the time axis of a moving reference frame cannot rotate beyond the 45 degree angle of light-speed.

4 Explaining Paradoxes

The Ladder and Barn Paradox

Recall from page 40 a ladder, slightly longer than a barn, can fit inside the barn when viewed by an observer in the barn because the ladder is seen to contract as it moves at vast speed. However, from the point of view of an observer moving with the ladder, the barn contracts and so the ladder cannot fit inside. How can this apparent paradox be explained away?

The phrase 'fits inside' actually means both ends of the ladder in the barn *at the same time*. Because of the Relativity of Simultaneity 'at the same time' means different things to the two observers because they are moving relative to each other, as we saw on page 110. For the observer in the barn, t_1 and t_2 are actually 'the same time' as far as she is concerned. But that is not the case for the observer travelling with the ladder. The ladder pokes out of one of of the contracted barn's doors (at t_1') then the other (at t_2') as each door closes in turn, and so there is no paradox.

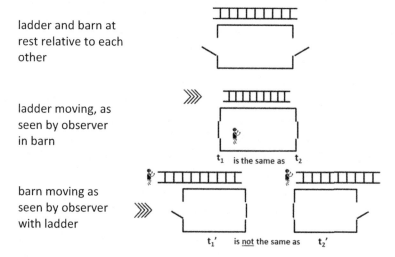

ladder and barn at rest relative to each other

ladder moving, as seen by observer in barn

t_1 is the same as t_2

barn moving as seen by observer with ladder

t_1' is not the same as t_2'

The Twin Paradox

We can understand the Twin Paradox using a spacetime diagram. Let's consider what's happening to the twins. Recall from page 40 that one of the twins travels away from the other at high speed, let's say to planet Zog one light-year away. There she turns around and returns. The apparent paradox is that since each twin is moving relative to the other, surely they both observe time slowing for the other twin. And so, when they reunite, each notices that the other has aged less than they!

Hidden inside the simple description of this seemingly persausive paradox is an inherent, but incorrect, presumption: that the situation of each twin is identical, i.e perfectly symmetrical. But as the spacetime diagram below highlights, the two situations are far from symmetrical.

The diagram is drawn from twin A's perspective. Each twin is in their own frame of reference. Measurements they make are made with respect to their own 'coordinate system'.

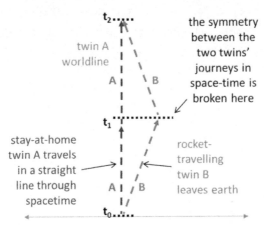

At t_1 twin B changes direction to return to twin A. We can see clearly from the diagram that the symmetry of the twins' experiences is broken at this time. This change in direction amounts to a change to her uniform motion. As such, according to the rules of special relativity, twin B is now in a different fame of reference, unlike twin A who remains in his. The next question is: how is that significant?

What will Twin A notice? His clock measures twin B's total journey. This agrees with his calculation (since he knows the distance to Zog and twin B's speed). However, he also observes slowing of twin B's clock, and therefore that twin B ages less than he.

What will Twin B notice? She of course believes her clock is ticking normally. But, because of length contraction, the distance to and from Zog appears shorter and she covers it more quickly: that is, in less time than has registered on twin A's clock. (This is exactly the same as the experience of the earth-bound muons in the relativity proof on page 68.) So she too observes that she has aged less than twin A. (Note that an identical conclusion will be reached even if the spacetime diagram is drawn from twin B's perspective instead.)

5 THE RELATIVITY FACTOR
(Also known as the Lorentz Factor)

Hendrik Lorentz was one of several physicists trying to make sense of the Michelson-Morley experimental result of 1887 that revealed the invariance of the speed of light (pages 28-29). And, before Einstein came along they'd arrived at a working mathematical description, albeit without having found a physical interpretation. There were two components. Firstly, the Relativity Factor that gave the amount by which distance, time and mass were affected. And secondly, rules (i.e. equations) which relate the perceptions of differently moving observers to each other; that is, how properties like distance, time and speed are **transformed between frames of reference**. (See pages 115-6 for more on transformations.)

Calculating the Relativity Factor
Consider the situation below noting the values: w, c, t, v, t' and d. An astronaut in the rocket will observe the light beam's vertical path, whether the rocket is moving or 'stationary', taking time t to traverse the rocket. However, when the rocket is moving a 'stationary' observer will observe the light take a longer diagonal path. Given light-speed invariance it will take longer to travel the greater distance, let's call that duration t'.

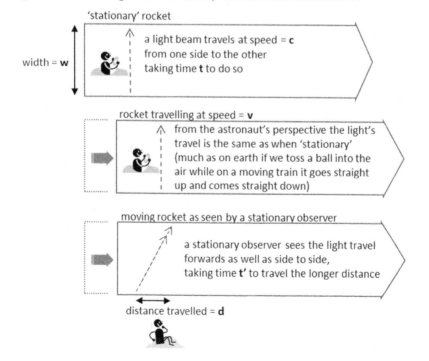

'stationary' rocket

a light beam travels at speed = **c**
from one side to the other
taking time **t** to do so

width = **w**

rocket travelling at speed = **v**

from the astronaut's perspective the light's travel is the same as when 'stationary' (much as on earth if we toss a ball into the air while on a moving train it goes straight up and comes straight down)

moving rocket as seen by a stationary observer

a stationary observer sees the light travel forwards as well as side to side, taking time **t'** to travel the longer distance

distance travelled = **d**

Now let's try to work out the relationship between t and t'.

From the astronaut's perspective, the distance travelled by the light is

w = c t   (1)   (from speed = distance divided by time)

From the 'stationary' observer's perspective, the distance travelled by the light is

c t'

But we know from Pythagoras' Theorem (describing the arithmetic relationship between the sides and diagonal of a right angle triangle) that

$$(ct')^2 = w^2 + d^2 \qquad (2)$$

And since we know the speed of the rocket, v, then we know that

d = v t'   (3)

So substituting equations (1) and (3) into equation (2) gives

$$(ct')^2 = (ct)^2 + (vt')^2$$

Dividing all three terms by c^2 gives

$$t'^2 = t^2 + (v/c)^2 \; t'^2$$

Rearranging

$$t^2 = t'^2 - (v/c)^2 \; t'^2$$
$$= t'^2 [1 - (v/c)^2]$$

So

$$t^2/t'^2 = [1 - (v/c)^2]$$

So

$$t/t' = sqrt [1 - (v/c)^2] \qquad \text{(NB sqrt = square root)}$$

Now replacing the right hand side by a mathematical abbreviation, γ, where

$$γ = 1 / sqrt [1 - (v/c)^2] \qquad = \text{Relativity Factor used to adjust t}$$

And so

$$t / t' = 1/γ$$

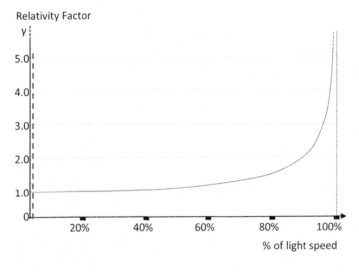

Relativity Factor

The equation draws the curve shown. Our everyday experience of speed lies to the left of the dashed line.

% of light speed

6 TRANSFORMATION OF COORDINATES AND SPEED BETWEEN MOVING FRAMES

Applying the Relativity Factor: Transformation of distance and time coordinates
We can use the Relativity Factor to calculate how the distance and time coordinates of an event at P observed by ('S') in his 'stationary' frame of reference can be transformed into coordinates in (M's) moving frame.

In this example we assume no motion along y or z axes, to simplify the workings out. However, similar equations can be derived for y and z.

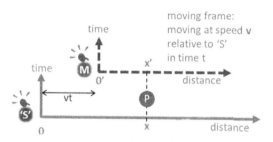

We derive the transformation for the distance coordinate as follows. In the 'S' frame the distance from the origin at 0 to P is x. In classical Newtonian physics to calculate the distance x' from 0' in the 'M' frame we would write
$$x' = x - v t \qquad (1)$$
In this equation vt is the distance the moving frame has travelled.

However, relativity says that length contracts by the Relativity Factor, γ. So
$$x' = \gamma (x - v t) \qquad (2)$$

We can derive the transformation for the time coordinate by analogy. (The actual derivation is long.) To start we expand the last term in equation (2) using c/c
$$x' = \gamma (x - (v/c) (ct))$$
this gives us two terms which we can describe as follows
  ct is a proxy measure of distance in the 'S' frame of reference (see page 117)
  v/c is a ratio which factors ct according to the M frame's speed

Then, by analogy, we arrive at a similar equation for t'
$$t' = \gamma (t - (v/c) (x/c))$$
where
  x/c is a proxy measure of time in the 'S' frame of reference
  v/c is the same ratio as above factoring x/c according to the M frame's speed
Reducing the previous equation we arrive at
$$t' = \gamma (t - v x/c^2) \qquad (3)$$

Inverse transformations can be derived in the same way to find results from the other observer's (M's) perspective, such that
$$x = \gamma (x' + vt') \qquad (4)$$
$$t = \gamma (t' + vx'/c^2) \qquad (5)$$

115

Applying the Relativity Factor: Transformation of speed; and Adding Speed
On page 41 we discussed how the combined speed of two objects could never exceed light-speed, no matter how fast each object appeared to be moving relative to a third 'stationary' observer. The following describes the maths behind that.

Consider a rocket moving at speed $v_R = 0.6c$ relative to a 'stationary' observer. Inside the rocket is an object moving in the same direction as the rocket at speed $u_R = 0.6c$ relative to the rocket. The question now is: how fast is the object moving relative to the 'stationary' observer? Let's call that u_O.

Newtonian physics (which works just fine at low speeds) would give the answer as the sum of the speeds
$$u_O = u_R + v_R = 0.6c + 0.6c = 1.2c$$
But relativity tells us that this is not possible, as it exceeds light-speed, c.

We arrive at the relativistic equation by considering the different reference frames of the observer and the rocket and making the appropriate transformations to distance and time. So instead of presuming that
$$u_O = x / t \qquad \text{(where x and t are distance and time in the rocket's frame)}$$
We recognise that
$$u_O = x' / t' \quad \text{(6)} \quad \text{(x' and t' are distance and time in the observers' frame)}$$

We wish to calculate the speed as seen from the observer's reference frame, so we substitute equations (4) and (5) from the previous page into (6) above
$$u_O = \frac{\gamma (x + v_R t)}{\gamma (t + v_R x/c^2)}$$
We can cancel out the γ terms. And dividing all remaining terms by t gives
$$u_O = \frac{x/t + v_R}{1 + v_R (x/t)/c^2}$$
$$= \frac{u_R + v_R}{1 + u_R v_R / c^2}$$
So in our example above
$$u_O = \frac{0.6c + 0.6c}{1 + (0.6c)(0.6c)/c^2} = 0.88c$$

Inverse transformations can be derived in the same way to find results from the other observer's perspective. So
$$u_R = \frac{u_O - v_R}{1 - u_O v_R / c^2}$$

'Proper', 'rest' or 'invariant' length is the length of an object as measured by an observer with the same motion as the object, i.e. at rest relative to each other. 'Proper' or 'invariant' time is measured by a clock with the same motion as an observer.

7 SPACETIME INTERVAL

The distance between two <u>points</u> in space

In 2D space, the distance, d, between two points is given by Pythagoras' Theorem.

$$d^2 = x^2 + y^2 \qquad \text{(x, y are the lengths of the two sides of a right-angle triangle)}$$

In 3D space, Pythagoras still applies.

$$d^2 = x^2 + y^2 + z^2 \qquad \text{(x, y, z are the distances between the two points in the three spatial dimensions)}$$

In the 3D geometry of classical physics if Newton observes a distance, d, in his frame of reference he can expect the same distance, d, will be measured by anyone else moving relative to his own reference frame. I.e. d is **'invariant'** in this context.

The distance between two <u>events</u> in spacetime

As we saw on page 45, in spacetime we don't talk about 'points' (i.e. 'where': x, y, x), we talk about 'events' (i.e. 'where' and 'when': x, y, z, t).

It's tempting to extend the above logic again to add a fourth dimension of time as: $+t^2$, but we can't. While space and time are related, they're not the same. We can't presume to simply extend Pythagoras into 4D spacetime. Geometry doesn't work the same here – why should it?

Furthermore, space and time have different units, metres and seconds. We measure the dimensions of space, x, y, z, in metres, and the dimension of time, t, in seconds. To combine 'distance in time' in the same equation as 'distance in space', we must somehow convert time to a proxy measure in metres. Since the speed of light, c, is constant, i.e. invariant across all frames of reference, it is safe to calculate this proxy measure as follows

'distance in time' = ct   (i.e. **c** *in metres/sec* times **t** *in secs* = **ct** *in metres*)

Furthermore, the word 'distance' loses meaning when applied to the time dimension. Instead, physicists talk about the **'spacetime interval'**. This conveys the idea of 'distance in space' and/or 'distance in time'. It is denoted by s^2, not s, for reasons we'll come to.

Imagine two events separated in space by an interval, d, where $d^2 = x^2 + y^2 + z^2$. The two events occur at different times, the interval in time being t (or ct once converted to our proxy quantity). We can presume that the spacetime interval, s^2, is a function of interval in space, d^2, and of interval in time, $(ct)^2$. Let's see what happens as the relative sizes of d^2 and $(ct)^2$ vary with respect to each other.

$d^2 > (ct)^2$ The 2 events are separated by more space than time. Nothing travels fast enough for one to **cause** the other: the interval is **'spacelike'**.

$d^2 < (ct)^2$ The 2 events are separated by more time than space. A signal can travel fast enough for one event to **cause** the other: the interval is **'timelike'**.

$d^2 = (ct)^2$ Only something travelling at the speed of light could be present at both at each event as it happens: the interval is **'lightlike'**.

These three cases correspond to areas of the spacetime diagram on page 109: the black region, the white region, and the dashed lines of the light cone's outline respectively.

The third equation, $d^2 = (ct)^2$ can help us arrive at the equation for the spacetime interval. For objects travelling at the speed of light distance has contracted to nothing and time has slowed to a standstill. The spacetime interval between two events in this case is 0. So we can rearrange the third equation

$$d^2 = (ct)^2$$

to give the result

$$d^2 - (ct)^2 = 0 = s^2$$

Or, as more normally written (expanding the term d^2)

$$s^2 = x^2 + y^2 + z^2 - (ct)^2$$
$$s^2 = x^2 - (ct)^2 \qquad \text{if y=0 and z=o} \qquad (1)$$

In this form, we can see why the spacetime interval is denoted by s^2 and not s. If the right hand side of the equation is expressed as a square root, it could result in a negative value, and physicists prefer to avoid unnecessarily having to take the squre root of a negative number.

Spacetime intervals are invariant

In our 3D Newtonian example, we saw that distance is invariant. It remains unchanged no matter the relative motion of any observer.

However, at high relativistic speeds, distance is not invariant. Differently moving observers will observe different lengths. And we know that, likewise, time is not invariant. However, in 4D spacetime, we find that the spacetime interval is invariant, i.e. the interval between two events as seen from the reference frames of two differently moving observers are equal. The proof is below (assumes y = 0 and z = 0 to simplify calculations).

From page 115 (assuming y=0 and z=0) we saw that

(2) $x' = \gamma (x - vt)$ &   (3) $t' = \gamma (t - vx/c^2)$
(4) $x = \gamma (x' + vt')$ &   (5) $t = \gamma (t' + vx'/c^2)$

So dividing (2) by (4) and (3) by (5) and cancelling γ

$x'/x = (x - vt)/(x' + vt')$ & $t'/t = (t - vx/c^2)/(t' + vx'/c^2)$

So, cross multiplying

$x'^2 + x'vt' = x^2 - xvt$ & $t'^2 + t'vx'/c^2 = t^2 - tvx/c^2$
Rearrange terms   *Now multiply this by c^2*
$x'^2 + t'vx' = x^2 - tvx$ & $c^2t'^2 + t'vx' = c^2t^2 - tvx$

Now subtract equation on right hand side from equation on left hand side

$x'^2 - c^2t'^2 = x^2 - c^2t^2$ (or $x'^2 - (ct')^2 = x^2 - (c^2t^2)$)

So, given (1) above,

$$s'^2 = s^2$$

So proving that spacetime intervals are invariant in 4D spacetime.

8 THE MATHS BEHIND INCREASING MASS

We described on pages 19 and 41 how an object's mass appears to increase as its speed increases. And we made clear that mass is not the same as matter. Rather, it is a measure of a property of matter called inertia: i.e. its resistance to change speed or direction.

A body with greater inertia has greater momentum too. Momentum is the amount of motion of a moving body. It is a measure of the tendency of a moving object to continue moving. For instance, the heavier our object or the faster its speed the more momentum it has: a fast-moving truck is harder to stop than a slow-moving bicycle.

So, in the form of an equation, momentum is a function of mass and speed

$$\rho = m v \qquad (\rho \text{ (Greek letter: rho)} = \text{momentum, } m = \text{mass, and } v = \text{speed})$$

$$\rho = m \frac{x}{t} \qquad (v = x/t = \text{distance/time. Assuming no travel along y or z axes})$$

That is the classical equation for momentum. Applying a relativistic perspective, the object is moving through an observer's reference frame. That observer will notice that time for the object, t', has slowed.

So if

$$\rho = m \frac{x}{t} \qquad (\text{where } t = \text{is the object's proper time})$$

Then for the observer the object's momentum is given by

$$\rho = \frac{m x}{t'} = \frac{m x}{t} \frac{t}{t'} \qquad (\text{i.e. expanding by multiplying by t/t})$$

But from time dilation we know that $t/t' = \gamma$ (see page 114). So

$$\rho = \gamma m v \qquad (\text{momentum as measured from stationary reference frame})$$

And since v has the same value as in the initial equation, the initial 'rest mass' m will appear to have increased by a factor of γ.

Misleading names

The term **'relativistic mass'** used to be used for this increased mass. The term **'rest mass'** (also known as **'proper mass'** or **'invariant mass'**) is still used to mean the mass of an object as measured by an observer who has the same motion as the object. Einstein himself did not like the term 'relativistic mass'. *'It is not good to introduce the concept of the [relativistic] mass of a moving body for which no clear definition can be given. It is better to introduce no other mass concept than the 'rest mass' ... [and] instead ... to mention the expression for the momentum and energy of a body in motion.'*

9 ARRIVING AT MASS-ENERGY EQUIVALENCE, $E=mc^2$

We saw on pages 49-51 how Einstein's famous equation highlights the enormous amount of energy locked up in matter.

Perhaps unsurprisingly given there are many forms of energy, there are a number of ways to derive the equation $E=mc^2$, two of which are included here. First is a very simple derivation. Second is one developed by Einstein himself, albeit four decades after he came up with the first version of $E=mc^2$ in 1905.

Very simple derivation of $E=mc^2$

Imagine an object travelling at vast speed, very nearly at light-speed. So close to light-speed in fact that any further force applied to the object produces a negligible increase in speed. As such we assume that all the energy from the extra force adds to the object's momentum as increased mass. Now imagine a force is applied to the object for one second.

From classical physics we know that
    E = F d   (E = energy, F = force, d = distance through which force applied)
Since the force was applied for 1 second and the object is travelling at very nearly light-speed, c, then we can assume that d = c. So we can write
    E = F c   (1)

From classical physics we also know that
    M = F t   (M = momentum gained, t = time for which force applied)
And also that
    M = m v   (m = mass of object, v = speed of object)
Combining these last two equations
    F t = m v
Since the force was applied for t=1 second and the object's speed, v, is very nearly c, so we can write
    F = m c   (2)

Combining (1) and (2) we get
    E = (m c) c
Or
    $E = m c^2$

(Note that this simple derivation works because we have ignored the negligible difference between the object's speed and light-speed, and that as a result all of the force converts to mass rather than speed. This sort of approximation is a frequently used mathematical practice.)

Einstein's 1946 derivation of E=mc²
As with the very simple derivation on the previous page, here Einstein relies on classical physics. This is the derivation which lies behind Thought Experiment #4 on page 49.

Imagine a box with mass m_0 floats freely in space. Two light sources are at rest relative to the box. At a given moment they both flash once, each striking the box with a light beam of energy E/2.

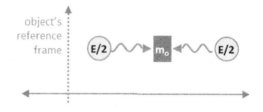

From Maxwell's equations electromagnetic waves have momentum (M_L), given by
   $M_L = E/c$ (energy divided by light-speed)
So in the situation above the total momentum received by the box is
   $M_L = 2 (E/2) /c$
   $= E/c$
However, since an equal momentum strikes from both sides, the box is unmoved.

Let's now look from the frame of reference of a rocket moving upwards at speed v. The rocket's astronaut considers himself 'stationary' and the box to be moving downwards at speed v.

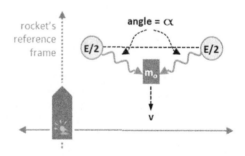

From Newton's equations the box's momentum before the light beams strike is
   $M_B = m_0 v$ (1)

The light beams strike the box at an angle.[27] The horizontal components of their momenta cancel out just as was the case in the box's reference frame. However, the vertical components of their momenta add to the box's vertical momentum.

From Maxwell's equation, again, the vertical component of the two light beams is
   $M_L = 2 [(E/2) \sin(\alpha)] / c$
   $= 2 [(E/2) (v/c)] / c$ ($\sin(\alpha) = v/c$ for very small angles)
   $= E v / c^2$ (2)

---

[27] Note that the light beam strikes the box at an angle because of the cosmological **'Aberration Effect'**. By analogy: despite vertically falling rain, the rain will strike a person walking fast through the rain at an angle.

The law of conservation of momentum says that momentum is conserved in collisions. So, in the astronaut's frame of reference after the light beams strikes the box's total momentum is

$$M_B = m_o v + E v/c^2$$

So

$$m'v = m_o v + E v/c^2 \quad \text{(where m' is the box's new, increased, mass)}$$
$$m' = m_o + E/c^2 \quad \text{(dividing both sides by v)}$$
$$E = (m' - m_o) c^2 \quad \text{(m'-m}_o\text{ = mass gained from the collision energy)}$$

So

$$E = m c^2 \quad \text{(where m = mass gained from light energy, i.e.}$$
$$\text{mass-energy equivalence of the light beams)}$$

Measurement units of $E=mc^2$

Energy (Joules) = Force x Distance through which force applied.

Energy E = [Force (Newtons)]   x [Distance (metres)]

= [Mass (kg) x Acceleration (metres/s^2)] x [Distance (metres)]

= kg x metres^2/s^2 which are same units as

= m c^2 (where 'm' means mass, not metres)

The mass-energy equivalence of a 100kg England male rugby player (See page 20.)

From $E=mc^2$ and c = 300,000,000 metres per second, so

1 kg of matter = 90,000 million million Joules of energy.

In 2012/3, the UK used 215 million tonnes of oil equivalent *

= 9,000,000 million million Joules of energy

So a 100 kg male rugby player would have serviced the total UK energy for a year if completely converted to energy through a nuclear reaction. **

* Source: www.gov.uk

** Note: a nuclear reaction only converts about 1% of matter to energy. This is the energy used to bind nuclear particles together. The remaining 99% is the energy from which those nuclear particles are made. (See page 51.)

10 Einstein's Other Famous Equation: General Relativity

Introduction

As far as the general population is concerned, Einstein's most famous equation is E=mc². But for physicists his most important equation is the one Stephen Hawking is talking about in his quote on page 107, the equation of General Relativity (GR). It is this equation which John Wheeler is describing in words when he says,

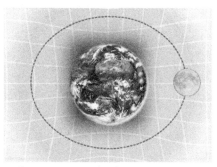

'Matter tells spacetime how to curve and curved spacetime tells matter how to move.'

The challenge then is how to describe in an equation: the curvature of the spacetime field[28]; its effect on objects; and the effect objects have on the spacetime field. The maths behind this equation is very complex, but here we'll try to shed a little light on its structure.

The essential idea can be described with an analogy. Hills and valleys can be represented on a 'topographical map' using 'contour lines' (as in the bottom of the two pictures below), each line being indexed with the height of the terrain. The closer the lines on the map, the steeper the terrain. By reading the map we can understand the size and shape of the hills and valleys, and the path a walker would take if he were intent on expending the least possible energy.

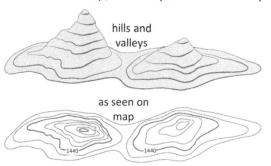

hills and valleys

as seen on map

Spacetime takes shape

Gravity causes particles to follow curved paths

On pages 53 and 54 we described how acceleration and gravity cannot be distinguished, and how the path of a light beam is seen to curve in some circumstances.

---

[28] Because the equation of General Relativity relates to the spacetime field it is often called '**Einstein's Field Equation**'.

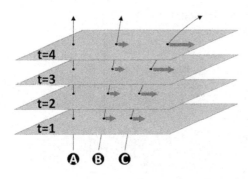

We can represent thought experiment #6 (on page 53) using our new tool of spacetime diagrams.

At left we see four slices of time, t=1 to t=4. Unlike the spacetime diagrams we saw earlier, each slice of time represents two of the three dimensions of space, not just one.

Three particles of light (A, B & C) have different trajectories, or worldlines.

- A considers itself 'stationary' within the reference frame of this spacetime diagram.
- B is moving at constant speed relative to A. So B pulls away from A by the same amount with each passing unit of time.
- C is gaining speed. So C pulls away from A by increasing amounts.

Special Relativity concerns itself with objects in uniform motion, resulting in straight line trajectories in spacetime, such as A and B above. General Relativity concerns itself with accelerating objects and their curved trajectories, such as C above.

Now consider four points on the earth, D, E, F & G. Because they are at different distances, the moon's gravity affects each point differently. (This is called the '**Tidal Effect**' as it gives rise to tides on earth.) The gravitational pull squashes the earth at D and E, and pulls F towards the moon. While G is pushed away because of the flattening at D and E.

As a result each point follows a different curved trajectory, as at right. And, the closer D, E and F come to the moon, the greater the effect on all four points.

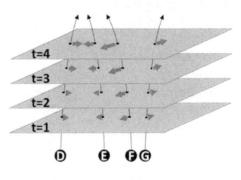

From curved paths to curved surfaces

We can see that particles D and E are accelerating towards each other, and so too do their curved worldlines. We can also see that the opposite is true of particles F and G. Therefore, a mathematical description of relativity for accelerating objects will need a way to represent these curved trajectories.

On a flat surface, the shortest distance between two points is a straight line. But on a curved surface the shortest distance is a curve called a '**geodesic**'. That's why an aircraft flying from Paris to Los Angeles travels over Greenland and not straight across the Atlantic, as we might expect from a two-dimensional map.

From this idea we see that we might be able to recreate the curved trajectories of particles D, E, F & G from the previous page using geodesics on curved surfaces. Looking at the three diagrams below, we can see that:

- A and B parallel the sides of a triangle drawn on a flat surface (so-called *zero curvature*). The triangle's interior angles add up to 180 degrees.
- D and E are curving towards each other, paralleling the sides of a triangle drawn on to a sphere. The triangle's interior angles add up to more than 180 degrees. (This is known as *positive curvature*.)
- F and G are curving away from each other, paralleling the sides of a triangle drawn on to a saddle. The triangle's interior angles add up to less than 180 degrees. (This is known as *negative curvature*.)

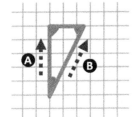 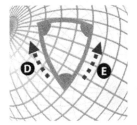 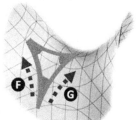

From drawings to equations

Let's start by teasing apart John Wheeler's quote, '*Matter tells spacetime how to curve and curved spacetime tells matter how to move*, and use this to provide a structure for the discussion which follows.

1. An object with mass …
2. is placed in 4D spacetime causing it to curve …
3. and this curvature in turn acts on other objects …
4. in a universe defined by constants such as the speed of light.

1 Describing an object's energy density, and the pressure it exerts
Matter is very densely packed energy. Physicists talk about 'energy density': referring either to low density energy forms (e.g. a photon) or high density forms (e.g. matter).

So while in Newton's equations we see 'mass' (m) appearing in his equations of motion, in Einstein's equation of GR we will see a term describing the energy density of objects.

> **Measurement units of energy density**
> Energy density is described as the amount of energy (measured in Joules) in a given volume, and is measured in Joules per metre cubed (J/m^3).

Recall also that the curved grid used to represent spacetime also represents the pressure gradients in the spacetime field. It is these pressure gradients which cause objects to move.

Energy density is closely related to pressure. For example, the air inside a balloon blown up to a certain *pressure* has an amount of stored *energy*. We see this energy in action when we let go of the balloon and it flies off.

> **Measurement units of pressure are same as those for energy density**
> Pressure is described as the amount of force per unit area, measured in (N/m^2), where Newtons, N, is the unit for force. Energy density and pressure are related. In the following [ ] mean 'measurement units of'.
>
> | [Pressure] | = [Force] | divided by | [Area] |
> |---|---|---|---|
> | | = [Force] x [Distance] | divided by | [Area] x [Distance] |
> | | = [Energy] | divided by | [Volume] |
> | | = [Energy density] | | |

2 Four-dimensional spacetime is curved by an energy density
Physicists don't know what spacetime is, nor the precise nature of the interaction between it and energy density. However, the effect of this interaction can be modelled and described mathematically.

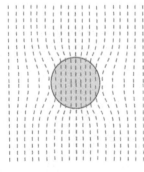

Physicists consider this interaction as analogous to a metal object in a magnetic field, see page 54. The object distorts the field, creating a concentration of field lines, and a local area of low pressure in the nearby magnetic field. As a result a pressure gradient is created between this and regions of higher pressure further away. These pressure gradients are like the gradients in the landscape of hills and valleys on page 123.

We understand that the bigger the metal object, the greater the distortion of the magnetic field. It is the same in General Relativity. The greater the energy density the greater the curvature of the spacetime field, for example, the black hole shown here.

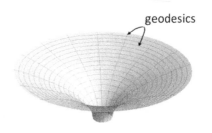

geodesics

3 Describing curved surfaces with maths
The degree of curvature of the solid line in the diagram changes along its length. We can draw circles which only just fit any curved region of the line. These are called 'osculating' (or kissing) circles. Depending on the degree of the line's curvature the radius (r) of the osculating circle varies.

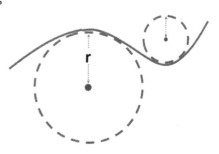

Measurement units of curvature
Mathematically, the degree of curvature is defined as the reciprocal of the radius, i.e. $1/r$. As the degree of curvature increases, the reciprocal of the radius of the osculating circle also increases. So ...
For curved lines
   In one dimension, such as the line above, the units are:
      $1/\text{length}$ or $1/\text{m}$ or m^{-1} (where m = metres)
For curved surfaces
   By inference, in two dimensions, such as a surface, the units are:
      $1/\text{area}$ or $1/\text{m}^2$ or m^{-2}

We also need a way to describe how the curve changes. At any point, P, on the curve a tangent can be drawn. We can imagine a force pulling that tangent down so that it follows the curve as P travels to the right along the curve.

As it does so, we can describe mathematically how much the tangent's slope changes in relation to the distance travelled by P. (That is, mathematically differentiating the tangent's slope with respect to distance travelled.)

4 A universe defined by constants

We have already learned of the important role the speed of light 'c' plays in the design of the universe. So it is not a surprise to find it in Einstein's equation of GR.

From school maths you may remember the important role that the mathematical constant π (pi) takes in any geometry involving curves. It is used when calculating for instance the circumference or area of a circle; or the surface area or volume of a sphere. So it's perhaps not a surprise, given the *curvature* of spacetime, that we will see it pop up also in Einstein's equation of GR.

The next constant we find in his equation is the **gravitational constant**, 'G'. This 'big G', as it is sometimes called, is the same as the big G in Newton's equations. The role played by this physical constant is simply to scale the answer that pops out of mathematical equations and thereby ensure the answers agree numerically with measured results.

Finally, there is the **cosmological constant** denoted by the Greek capital letter lambda (Λ). This is discussed in more detail on the next page.

Constant values and their units
c = 299,792,458 m/s   (m = metres; s = seconds)
π = 3.14159...   (pi has no units)
G = 6.674×10^{-11} (N/kg^{-2}) m^2 (N = Newtons; kg = kilograms)
Λ = [Value depends! See next page] m^{-2}

Einstein's equation of General Relativity

Einstein's equation of General Relativity can be written in different ways which are all, of course, in essence the same. This is a common form:

$$R_{\mu v} - \frac{R\, g_{\mu v}}{2} + \Lambda\, g_{\mu v} = \frac{8\,\pi\,G\,T_{\mu v}}{c^4}$$

For a region of spacetime, the equation's left hand side describes its curved geometry, and the right hand side its energy density. As such it relates spacetime geometry to energy density. The terms $R_{\mu v}$ $g_{\mu v}$ and $T_{\mu v}$ are known as '**tensors**'. The terms are explained on the following pages.

Measurement units of GR equation
The terms to the left of the equals sign are all $1/m^2$ (or m^{-2}) relating to the measurement of curvature. To the right, the term $T_{\mu v}$ is in J/m^3 relating to energy density. (Both sets of units were discussed earlier.) Once the various constants (π, G and c) are applied to $T_{\mu v}$ the units on the right hand side match the units on the left.

$R_{\mu\nu}$, $g_{\mu\nu}$ and $T_{\mu\nu}$ are 'tensors'. The subscript characters 'μ' (Greek letter 'mu') and 'v' (Greek letter 'nu') represent the four dimensions of spacetime.
Conceptually, tensors are similar to vectors. (Vectors are actually a simple type of tensor.) Vectors are used to represent the size and the direction of, among other things: (1) the speed of an object in 'normal' space, and (2) force. However, more complex terms are needed to represent similar but more complex concepts such as: (1) the path of an object in 4D curved spacetime, and (2) the stress induced in a region of spacetime by energy density where force acts in more than one direction at the same time, much like the pressures caused by an object in water. Two subscript letters are used to describe the components of tensors in the four dimensions of spacetime, μ and v. Each can have four values, conventionally: 0 for the time dimension and 1, 2, 3 for the space dimensions. The subscripts tell us: don't forget to add together the components across all of the dimensions. That's why the GR equation is described as a set of 10 equations. Since μ and v can each have 4 values, there are 2^4 =16 possible combinations of values. However, 6 of these are mathematically equivalent to others (e.g. μ=1 and v=2 is equivalent to μ=2 and v=1) so that leaves 10 unique equations in the set.

The terms '$R_{\mu\nu}$' (Ricci tensor) and 'R' (Ricci scalar) describe spacetime's shape.
Together they describe the complex 4D curved geometry of a region of spacetime.

The term '$g_{\mu\nu}$' (metric tensor) captures the causal structure of spacetime.
It describes which events in spacetime (i.e. space _and_ time) can influence which other events. As such, it comprises notions such as time, distance, volume, curvature, angle and separation of the future and the past. (See the discussion on page 109 regarding causality in regions of spacetime.)

The term '$T_{\mu\nu}$' is the stress-energy-momentum tensor.
This describes the energy density at a particular location in spacetime in terms of the quantity of energy and momentum and in which direction it is flowing.

Cosmological constant lambda (Λ), Dark energy, and Expanding universe

Traditionally, Λ has been a constant relating to curvature, hence its presence to the left of the equals sign. Einstein introduced the constant so that his equation described a static universe (not expanding or contracting).

Once it was confirmed that the universe was in fact expanding, he set Λ to zero. In 1998, it was discovered that the **universe's expansion** was accelerating. This requires Λ to have a small positive value.

It is speculated that this expansion is fuelled by some **dark energy** which would therefore have an energy density. It is mathematically feasible to move the Λ term to the right of the equals sign to join with the other term for energy density, hence its relevance to dark energy.

The shape of the universe

On page 59 we said that the current opinion of cosmologists is that the universe, on average, is **not curved but 'flat'**. Using the ideas discussed on page 125 we can now better understand what that means.

In places the universe will have zero curvature, in others positive or negative curvature. But the mathematical average of these hills and valleys in spacetime is believed to be roughly zero. As we saw, flat surfaces are described as having zero curvature. That's why at times the universe is described as 'flat'. But it is not the same sort of flatness we associate with the everyday use of the word 'flat'.

Also on page 59 we said that the universe (that is 4D spacetime) was **finite but unbounded**. The simplest way to visualise what this means is by analogy to a 2D-thinking alien living on a 3D sphere. You and I know that the sphere has a *finite* volume. But the 2D-thinking alien has no understanding of the sphere or that he could somehow leave its surface. To the alien, the surface is *unbounded*, in that he can walk forever in any direction and not reach any edges at all. Eventually, he might discover that he has come back to where he started. The same could theoretically be true for us in 4D spacetime!

Einstein and Newton

It is helpful to consider how Newton's and Einstein's theories relate to each other. After all Newton's equations work perfectly well at low relative speeds of travel. So it is no surprise that at low speeds, Einstein's equation of GR can be simplified mathematically to produce Newton's equations. (How this simplification happens is beyond the scope of this book.)

We understand that in Einstein's relativity:
    Mass-energy tells spacetime how to curve; and
    curvature tells mass-energy how to move.

Whereas in Newton's universe:
    Mass determines the force of gravity;
$$F = G\, m_1 m_2 / r^2$$
    and, force tells mass how to accelerate.
$$F = m\, a$$
    (where F is the force of attraction between two objects with masses m_1 and m_2 separated by a distance r; G is the same Gravitational Constant 'Big G' as in Einstein's equation; and, a is acceleration.)

ACKNOWLEDGEMENTS

I did not set out to write a book, only to understand relativity. It was not the first time. There had been many failed attempts. The idea 'that the universe was not as it seemed' compelled perseverance.

Thanks to many generous inhabitants of the internet there are several good explanations of relativity which, while still fairly technical, were partially comprehensible even to me. These and other sources are listed at the end of this book and I am indebted to them all.

Finally, early in 2013, with progress being made I began to write notes, lest I forget! So this book was born. My first draft, perhaps unsurprisingly, had significant errors. The late Dr Rodney Hillier, Bristol University Astrophysics Department Emeritus, was patiently generous with his time in explaining where these were. This was typical of a man who, even after retirement, gave readily to the public advancement of science.

A number of family and friends, many of whom are not of scientific bent let alone shown any previous interest in relativity, read later drafts and picked me up whenever I succumbed to not writing simply or clearly. So thanks to Henri, Jo, Katharine and David Hulman, to John Gordon and Tom Sheppey. And, especial thanks to my wife Sarah, who, despite possibly being the least interested initially, read two drafts and challenged my explanations repeatedly and very helpfully. (I'm sorry you ended up dreaming about relativity.)

Finally, I am particularly grateful to Professor Elias Brinks of the Centre for Astrophysics Research at the University of Hertfordshire. He responded to my unsolicited request for help to check the manuscript. Despite a high research and teaching load, he reviewed the final drafts and helped me understand and correct the more nuanced aspects of relativity. He also recruited others to review the draft, all of whom provided helpful comments: Jake Dungworth, James Hanson, Rod Martin and Lewis Palmer. My sincere thanks to all these also. From the outset, Professor Brinks kept faith with the book's intended simplicity, guiding me through final corrections with a style of teaching I imagine his students must appreciate greatly.

Final responsibility cannot, of course, be avoided. Any remaining faults are mine and mine alone.

REVISION HISTORY

First published October 2016

NOTE Professor Brinks has not reviewed the following changes.
Revisions April 2018
- Chapter *Relativity Over a Cup of Coffee*
- Pages 40-45 of chapter *More Detailed Logic*
- Additions to the chapter *A Little Bit of Maths*: Parts 3, 6 and 7

Revisions January 2019
Revisions to 'Apparently increasing mass' in both *Relativity Over a Cup of Coffee* and *More Detailed Logic*, and Part 8 of *A Little Bit of Maths*
Revisions June 2021
- Preface added
- Chapter *More Detailed Logic* pages 31, 49-51, 53
- Chapter *Theory of Everything, Quantum Matters* pages 81-82
- Chapter *On 'Time'*
- Chapter *A Little Bit of Maths*
    Pages 108, 111
    Part 9 pages 120-122
    Part 10 pages 123-130

Revisions July 2023
- Chapter *Relativity Over a Cup* of Coffee pages 13-18

At various times
Amendments and corrections.

BIBLIOGRAPHY

CHAPTERS ON RELATIVITY: CUP OF COFFEE, DETAILED LOGIC, PROOF, A LITTLE BIT OF MATHS

Before Einstein: Mechanical explanations of gravitation
  en.wikipedia.org/wiki/Mechanical_explanations_of_gravitation
Relativity, summary explanations
  Einstein Person of the Century, Time Magazine December 31st 1999
  Relativity: Einstein's Mind-Bending Universe, New Scientist Collection Volume 4 Issue 1
  General Relativity. The most beautiful theory. The Economist, November, 28th 2015
  [Audio] *In Our Time – Relativity* bbc.co.uk/programmes/b02144gl
  The Ascent Of Man, pp245-258, by J Bronowski
  Ask Ethan series of articles in Forbes Magazine, by Ethan Siegel
  The Frontiers of Knowledge, pp78-140, by AC Grayling
Relativity, detailed explanations
  phys.unsw.edu.au/einsteinlight/
  einstein-online.info/
  physicsoftheuniverse.com/topics_relativity.html
  physics.nyu.edu/hogg/sr/sr.pdf
  www.phys.ufl.edu/~acosta/phy2061/lectures/Relativity4.pdf
  http://hyperphysics.phy-astr.gsu.edu/hbase/hframe.html
  einstein.stanford.edu/SPACETIME/spacetime-index.html
  www1.phys.vt.edu/~takeuchi/relativity/notes/index.html
  www.pitt.edu/~jdnorton/teaching/HPS_0410/index.html
  www.khanacademy.org/science/physics/special-relativity
  phys.libretexts.org/ ... The_Theory_of_Special_Relativity
Einstein's first paper on relativity fourmilab.ch/etexts/einstein/specrel/www/
Invariance of the speed of light & Principle of relativity
    physicsoftheuniverse.com/topics_relativity_light.html
Michelson-Morley Experiment en.wikipedia.org/wiki/Luminiferous_aether
Special Relativity and Spacetime Interval
  en.wikipedia.org/wiki/Special_relativity
  en.wikipedia.org/wiki/Spacetime
  www.webassign.net/question_assets/buelemphys1/chapter26/section26dash2.pdf
  www.feynmanlectures.caltech.edu/I_15.html
Spacetime geometry of length contraction & time dilation (slowing of time)
  [Animation] youtube.com/watch?v=lVuF5zrwMLY
  en.wikipedia.org/wiki/Length_contraction
Visual appearance of length contraction and time dilation
  en.wikipedia.org/wiki/Terrell_rotation
The Twin Paradox
  en.wikipedia.org/wiki/Twin_paradox
  newt.phys.unsw.edu.au/einsteinlight/jw/module4_twin_paradox.htm
Muon experiment, Relativity proof hyperphysics.phy-astr.gsu.edu/hbase/Relativ/muon.html
Photons [Audio] *In Our Time – The Photon* bbc.co.uk/podcasts/series/iot#playepisode5
The momentum and mass of light
  van.physics.illinois.edu/qa/listing.php?id=1424
  physics.stackexchange.com/... why-does-the-mass-of-an-object-increase
Faster-than-light travel en.wikipedia.org/wiki/Faster-than-light

What is spacetime made of?
    abc.net.au/science/articles/ ... greatmomentsinscience
    blogs.scientificamerican.com/ ... what-do-you-mean-the-universe-is-flat
    en.m.wikipedia.org/wiki/Spacetime
    www.forbes.com/sites/startswithabang/2018/08/11/ask-ethan-is-spacetime-really-a-fabric/
Expanding universe / Big Bang
    crystalinks.com/bigbang.html
    io9.com/5881330/what-happened-before-the-big-bang
    skyserver.sdss.org/dr1/en/astro/universe/universe.asp
    https://cosmology.carnegiescience.edu/timeline.html
    https://www.forbes.com/sites/.../what-came-first-inflation-or-the-big-bang/?sh=3d38093e4153
The derivation of $E=mc^2$
    The Sextant Equation: $E=mc^2$, by Peter Galison
        https://galison.scholar.harvard.edu/publications/sextant-equation-emc2
    adamauton.com/warp/emc2.html
Relativistic momentum
    www.feynmanlectures.caltech.edu/I_16.html
    real-world-physics-problems ... difference-between-momentum-and-kinetic-energy
The equivalence principle
    en.wikipedia.org/wiki/Equivalence_principle
    phys.org/news/2011-11-mass-energy.amp
The equation of General Relativity
    en.wikibooks.org/wiki/General_Relativity/Einstein%27s_equation
    www.physicsforums.com/threads/units-of-cosmological-constant.708526/
    Tensors https://www.youtube.com/watch?v=f5liqUk0ZTw
History of relativity
    en.wikipedia.org/wiki/History_of_special_relativity
    www.wikiwand.com/en/History_of_special_relativity#/overview
    Influence of Hume and Mach www.pitt.edu/~jdnorton/papers/HumeMach.pdf
    Clock tower thought experiment, myth or truth? *Science Secrets*, by Martinez
    Did Einstein know about the Michelson-Morley experiment?
        adsabs.harvard.edu/full/1974Obs....94...81J
        physics.stackexchange.com/.../did-einstein-know-about-the-michelson-morley-experiment
    Einstein letter to *The Times* germanhistorydocs.ghi-dc.org/pdf/eng/EDU_Einstein_ENGLISH.pdf

CHAPTER: THEORY OF EVERYTHING, QUANTUM MATTERS
Quantum mechanics
    The Quantum Universe: Everything that can happen does happen, by Cox & Forshaw
    50 Quantum physics ideas you really need to know, by Joanne Baker
    Quantum Theory Cannot Hurt You, by Marcus Chown
    In Search of Schrödinger's Cat, by John Cribben
    The Quantum Self, pp1-32, by Danah Zohar
    The Fabric of Reality, pp1-53, by David Deutsch
    New Scientist, Essential Guide No. 5 Quantum Physics
    Reality Is Not What It Seems, by Carlo Rovelli
    On the Origin of Time, Thomas Hertog
    https://wtamu.edu/ ... why-do-quantum-effects-only-happen-on-the-atomic-scale/
Wavefunction www.khanacademy.org/science/physics/.../v/quantum-wavefunction
Uncertainty principle https://scienceexchange.caltech.edu/topics/.../uncertainty-principle

Quantum entanglement https://www.sciencedaily.com/terms/quantum_entanglement.htm

EPR Paradox: *The EPR Paper & Bell's Theorem*, by Mikey Pilger

    www3.nd.edu/.../Pilger%20The%20EPR%20Paper%20and%20Bells%20Theorem.pptx

Wigner's Friend www.scientificamerican.com/article/this-twist-on-schroedingers-cat-paradox-has-major-implications-for-quantum-theory/

Standard Model

    physics.info/standard/

    https://home.cern/science/physics/standard-model

CHAPTER: ON 'TIME'

Time

    Seven Brief Lessons on Physics pp49-62, by Carlo Rovelli

    www.exactlywhatistime.com

    www.quantamagazine.org ... does-time-really-flow

    The arrow of time (why time only flows forwards)

        preposterousuniverse.com/eternitytohere/faq

        informationphilosopher.com/problems/arrow_of_time/

CHAPTER: RELATIVITY'S PHILOSOPHICAL IMPLICATIONS

Perception of dimensions

    Flatland: A Romance of Many Dimensions, by Edwin A. Abbott

Exploring implications of modern physics

    The Mind of God, by Paul Davies

Quantum weirdness: The battle for the basis of reality

    currenteventsii.yuku.com/topic/21231

CHAPTER: RELATIVITY, CULTURE AND SOCIETY

General

    Einstein for the 21st Century: His Legacy in Science, Art, & Modern Culture, Peter Galison

    sparknotes.com/biography/einstein/section12.rhtml

    Einstein Person of the Century, Time Magazine December 31st 1999

Atomic bomb

    The Making of the Atomic Bomb, by Richard Rhodes

    amnh.org/exhibitions/past-exhibitions/einstein/peace-and-war/the-manhattan-project

Einstein's philosophy of science plato.stanford.edu/entries/einstein-philscience/

Art

    scq.ubc.ca/the-art-of-science-the-science-of-art/

    theguardian.com/artanddesign/2004/may/22/art

    [Audio] *Start the Week – Anish Kapoor, Light & Dark* bbc.co.uk/programmes/b0785nl7

Anti-semitism

    scientificamerican.com/ ... pro-nazi-nobelists-attacked-einstein

    en.wikipedia.org/wiki/Deutsche_Physik

    arxiv.org/ftp/arxiv/papers/1111/1111.2194.pdf

CHAPTER: MORE ABOUT EINSTEIN

Does Einstein deserve the credit?

    en.wikipedia.org/wiki/Relativity_priority_dispute

    naturelovesmath.com/en/ ... einstein-vs-poincare

    physicscentral.com/explore/writers/rigden.cfm

    scientificamerican.com/article/was-einstein-the-first-to-invent-e-mc2/

Popular opposition to relativity and to Einstein. Einstein's Nobel Prize.

    guardian.co.uk/science/across-the-universe/2012/oct/08/einstein-nobel-prize-relativity

PERMISSIONS

QUOTES
Quotes in this book are presumed 'Fair Use'.

GRAPHICS
Images or graphics not listed below are copyright of the author.

IMAGES
Permissions granted.

Page

2. Adapted from HubbleSite with permission from NASA and STScI
8. From '*Life*' (c) Edward Monkton. Reproduced with kind permission of the artist.
69. *Relativity Theory* (c) Chris Madden. Reproduced with kind permission of the artist.

Images free to use or presumed 'Fair Use'.

Page

Cover. Clock, by Gerd Altmann. From Pixabay (279334)
1, 65, 66. From Milky Way, by Skeeze. From Pixabay (916523)
3. Pale Blue Dot. NASA/JPL-Caltech
9. Reykjavik Cathedral, Iceland, by Unsplash. From Pixabay (569340)
10. Ireland, fields, sky, clouds, by Jon Sullivan. From public-domain-image.com
   South Africa Mesa Nature Hills, by Life of Pix. From Pixabay (406914)
20. Rugby player (Martin Johnson). Wikipedia
21. Spacetime, pbslearningmedia.org, 'Curvature-of-space-light-falls'
27. Ships, by Rujhan Basir Pixabay (1719396)
28. Adapted from graphic of Electromagnetic wave. Several sources. Origin unknown.
39. Rainbow Row, Charleston. From Wikipedia
54. Magnetic field, Several sources
54. Spacetime Christopher Vitale of Networkologies and The Pratt Institute
56. Black hole, by Chris McDonnell, thenounproject.com (8472)
58. Adapted from Gravitational wave. Source: spaceplace.nasa.gov/review/lisa-g-waves/
58. Reflection, lake in Ditchingham Park. From Wikimedia Commons.
63. Einstein cartoon. Amazing Formula, by Javad Alizadeh. From Wikimedia Commons.
72. Water ripples. From stopdown.net/interference.htm
74. Interference 1. From wired.com/2014/06/the-new-quantum-reality/
74. Interference 2. From thestargarden.co.uk/Quantum-mechanics.html
79. Atom. From writescience.wordpress.com/2014/09/27/science-folk-tales/
80. Rocky Mountain, by Unsplash. From Pixabay (1149298)
85. Clock, by Gerd Altmann. From Pixabay (598837)
96. Monkey Typing Shakespeare, by Andand S., flickr (3379486150)
96. Adapted from NASA image and a section of Michelangelo's 'The Creation of Adam'
99. *Relativity* by MC Escher. 'Fair Use' reproduction (as per Wikipedia).
101. Einstein in 1904. Wikipedia
107. Einstein in 1921, by F Schmutzer. From Wikimedia Commons.
123. Topgraphy, www.rei.com/learn/expert-advice/topo-maps-how-to-use.html
127. 'Black hole.' Adapted from Flamm's paraboloid. From Wikipedia

INDEX